Lecture Notes in Artificial Intelli

Edited by R. Goebel, J. Siekmann, and W. Wa

Subseries of Lecture Notes in Computer Science

Michael Fisher   Fariba Sadri
Michael Thielscher (Eds.)

Computational Logic
in Multi-Agent Systems

9th International Workshop, CLIMA IX
Dresden, Germany, September 29-30, 2008
Revised Selected and Invited Papers

 Springer

Series Editors

Randy Goebel, University of Alberta, Edmonton, Canada
Jörg Siekmann, University of Saarland, Saarbrücken, Germany
Wolfgang Wahlster, DFKI and University of Saarland, Saarbrücken, Germany

Volume Editors

Michael Fisher
University of Liverpool
Department of Computer Science
Ashton Street, Liverpool, L69 3BX, UK
E-mail: MFisher@liverpool.ac.uk

Fariba Sadri
Imperial College London
Department of Computing
180 Queen's Gate, London, SW7 2AZ, UK
E-mail: fs@doc.ic.ac.uk

Michael Thielscher
Dresden University of Technology
Department of Computer Science
Artificial Intelligence Institute
Computational Logic Group
Nöthnitzer Str. 46, 01187 Dresden, Germany
E-mail: mit@inf.tu-dresden.de

Library of Congress Control Number: 2009930215

CR Subject Classification (1998): F.3, D.3.2, F.4, I.2.3-4, D.3, H.1.1, G.2

LNCS Sublibrary: SL 7 – Artificial Intelligence

ISSN   0302-9743
ISBN-10   3-642-02733-4 Springer Berlin Heidelberg New York
ISBN-13   978-3-642-02733-8 Springer Berlin Heidelberg New York

springer.com

© Springer-Verlag Berlin Heidelberg 2009
Printed in Germany

Typesetting: Camera-ready by author, data conversion by Scientific Publishing Services, Chennai, India
Printed on acid-free paper   SPIN: 12693702   06/3180   5 4 3 2 1 0

Preface

Multi-Agent Systems are communities of problem-solving entities that can exhibit varying degrees of intelligence. They can perceive and react to their environment, they can have individual or joint goals, for which they can plan and execute actions. Work on such systems integrates many technologies and concepts in artificial intelligence and other areas of computing as well as other disciplines. The agent paradigm has become very popular and widely used in recent years, due to its applicability to a large range of domains, from search engines to educational aids, to electronic commerce and trade, e-procurement, recommendation systems, and ambient intelligence, to cite only some.

Computational logic provides a well-defined, general, and rigorous framework for studying syntax, semantics and procedures for various capabilities and functionalities of individual agents, as well as interaction amongst agents in multi-agent systems. It also provides a well-defined and rigorous framework for implementations, environments, tools, and standards, and for linking together specification and verification of properties of individual agents and multi-agent systems.

The CLIMA workshop series was founded to provide a forum for discussing, presenting and promoting computational logic-based approaches in the design, development, analysis and application of multi-agent systems. The first workshop in this series took place in 1999 in Las Cruces, New Mexico, USA, under the title Multi-Agent Systems in Logic Programming (MASLP 1999), and was affiliated with ICLP 1999. The name of the workshop changed after that to Computational Logic in Multi-Agent Systems (CLIMA), and it has since been held in countries including UK, Cyprus, Denmark, USA, Portugal, Japan and Germany. Further information about the CLIMA series, including past and future events and publications can be found at http://centria.di.fct.unl.pt/~clima.

The ninth edition of CLIMA (CLIMA IX) was held on 29-30 September 2008 in Dresden, Germany. It was co-located with the 11th European Conference on Logics in Artificial Intelligence (JELIA 2008). More details about the event can be found at http://www.csc.liv.ac.uk/~michael/clima08.html.

This volume of proceedings contains revised and improved versions of eight of the workshop regular papers and two invited contributions, one a full paper by Wojciech Jamroga, and another, an extended abstract by Mehdi Dastani. All the papers included in the proceedings have gone through a thorough revision process, with at least two rounds of reviewing. The papers cover a broad range of topics. The invited contributions discuss complexity results for model-checking temporal and strategic properties of multi-agent systems, and the challenges in design and development of programming languages for multi-agent systems.

The topics discussed in the regular papers include the use of automata-based techniques for verifying agents' conformance with protocols, and an approach based on the C+ action description language to provide formal specifications of social processes such as those used in business processes and social networks. Other topics include casting reasoning as planning and thus providing an analysis of reasoning with resource bounds, a discussion of the formal properties of Computational Tree Logic (CTL) extended with knowledge operators, and the use of argumentation in multi-agent negotiation.

We would like to thank all the authors for responding to the call for papers with their high quality submissions, and for revising their contributions for inclusion in this volume. We are also grateful to the members of the CLIMA IX 2008 Program Committee and other reviewers for their valuable work in reviewing and ensuring the quality of the accepted papers. We would also like to thank the local organizers in Dresden for all their help and support. We are grateful to them for handling all the registration details and providing an enjoyable social program.

April 2009

Michael Fisher
Fariba Sadri
Michael Thielscher

Organization

CLIMA Steering Committee

Jürgen Dix — Technical University of Clausthal, Germany
Michael Fisher — University of Liverpool, UK
João Leite — New University of Lisbon, Portugal
Fariba Sadri — Imperial College London, UK
Ken Satoh — National Institute of Informatics, Japan
Francesca Toni — Imperial College London, UK
Paolo Torroni — University of Bologna, Italy

CLIMA IX 2008 Program Committee

Natasha Alechina — University of Nottingham, UK
Jose Julio Alferes — New University of Lisbon, Portugal
Rafael H. Bordini — University of Durham, UK
Gerhard Brewka — Leipzig University, Germany
Stefania Costantini — University of Aquila, Italy
Mehdi Dastani — Utrecht University, The Netherlands
Marina De Vos — University of Bath, UK
Juergen Dix — Clausthal University of Technology, Germany
Michael Fisher — University of Liverpool, UK
Chiara Ghidini — FBK irst, Italy
James Harland — RMIT University, Australia
Hisashi Hayashi — Toshiba, Japan
Katsumi Inoue — National Institute of Informatics, Japan
Joao Leite — New University of Lisbon, Portugal
Fangzhen Lin — Hong Kong University of Science and Technology, Hong Kong
Viviana Mascardi — University of Genoa, Italy
Paola Mello — University of Bologna, Italy
John Jules Meyer — Utrecht University, The Netherlands
Leora Morgenstern — IBM T.J. Watson Research Center, USA
Naoyuki Nide — Nara Women's University, Japan
Mehmet Orgun — Macquarie University, Australia
Maurice Pagnucco — The University of New South Wales, Australia
Jeremy Pitt — Imperial College London, UK
Enrico Pontelli — New Mexico State University, USA
Fariba Sadri — Imperial College London, UK
Chiaki Sakama — Wakayama University, Japan
Ken Satoh — National Institute of Informatics, Japan
Renate Schmidt — The University of Manchester, UK

Tran Cao Son	New Mexico State University, USA
Michael Thielscher	Dresden University of Technology, Germany
Francesca Toni	Imperial College London, UK
Wiebe van der Hoek	University of Liverpool, UK
Cees Witteveen	Delft University of Technology, Netherlands

CLIMA IX 2008 Additional Reviewers

Francisco Azevedo
Tristan Behrens
Naoki Fukuta
Wojtek Jamroga
Jianmin Ji
Naoyuki Nide
Peter Novak
Dmitry Tishkovsky
Yisong Wang

Table of Contents

Easy Yet Hard:
Model Checking Strategies of Agents

Wojciech Jamroga

Department of Informatics, Clausthal University of Technology, Germany
wjamroga@in.tu-clausthal.de

Abstract. I present an overview of complexity results for model checking of temporal and strategic logics. Unfortunately, it is possible to manipulate the context so that different complexity results are obtained for the same problem. Among other things, this means that the results are often distant from the "practical" complexity which is encountered when one tries to use the formalisms in reality.

1 Introduction

A study of computational complexity is nowadays almost obligatory in a paper on logic in AI. Authors usually study the complexity of model checking and/or satisfiability checking of their logic in order to back the usefulness of the proposal with a formal argument. Unfortunately, the results are often far from the "practical" complexity which is encountered when one tries to use the formalisms in reality. Moreover, it is possible to manipulate the context so that different complexity results are obtained for the same problem. In this paper, I present a brief overview of complexity results for model checking temporal and strategic logics. Three logics are discussed here, namely computation tree logic CTL, alternating-time temporal logic ATL, and alternating-time logic with imperfect information and imperfect recall ATL$_{ir}$. For these logics, I show how the complexity class of the model checking problem changes when we change the way we represent input and/or measure its size.

Does it mean that theoretical complexity results are not worth anything in practice? Not necessarily – but certainly one needs to take these results with a grain of salt. In most cases, only a more extensive study (carried out from several different perspectives) can give us a meaningful picture of the *real* computational difficulty behind the problem.

2 The Logics

2.1 CTL: Branching Time and Temporal Evolution

Computation tree logic CTL [4,6] explicitly refers to patterns of properties that can occur along a particular temporal path, as well as to the set of possible time series. The first dimension is captured by *temporal operators*: "$\bigcirc$" ("in the

M. Fisher, F. Sadri, and M. Thielscher (Eds.): CLIMA IX, LNAI 5405, pp. 1–12, 2009.
© Springer-Verlag Berlin Heidelberg 2009

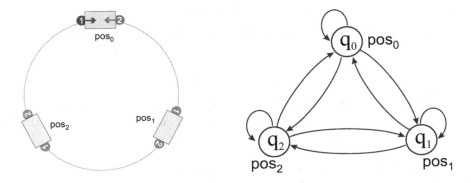

Fig. 1. Two robots and a carriage: a schematic view (left) and a transition system M_0 that models the scenario (right)

next state"), $\square$ ("always from now on") and $\mathcal{U}$ ("until"). Additional operator $\diamond$ ("sometime from now on") can be defined as $\diamond \varphi \equiv \top \mathcal{U} \varphi$. The second dimension is handled by so called *path quantifiers*: E ("there is a path") and A ("for all paths"). In CTL, every occurrence of a temporal operator is preceded by exactly one path quantifier.[1] Formally, the recursive definition of CTL formulae is:

$$\varphi ::= p \mid \neg \varphi \mid \varphi \wedge \varphi \mid \mathsf{E} \bigcirc \varphi \mid \mathsf{E} \square \varphi \mid \mathsf{E} \varphi \mathcal{U} \varphi.$$

A is derived from E in the usual way (cf., e.g., [13]).

The semantics of CTL is defined over *unlabeled transition systems*, i.e., tuples $M = \langle St, \mathcal{R}, \Pi, \pi \rangle$ where St is a nonempty set of states (or possible worlds), $\mathcal{R} \subseteq St \times St$ is a serial transition relation on states, Π is a set of atomic propositions, and $\pi : \Pi \rightarrow 2^{St}$ is a valuation of propositions. A *path* (or *computation*) in M is an infinite sequence of states that can result from subsequent transitions, and refers to a possible course of action. Let $\lambda[i]$ denote the ith position in computation λ (starting from $i = 0$). The meaning of CTL formulae is given by the following clauses:

$M, q \models p$ iff $q \in \pi(p)$ (where $p \in \Pi$);
$M, q \models \neg \varphi$ iff $M, q \not\models \varphi$;
$M, q \models \varphi \wedge \psi$ iff $M, q \models \varphi$ and $M, q \models \psi$;
$M, q \models \mathsf{E} \bigcirc \varphi$ iff there is a path λ such that $\lambda[0] = q$ and $M, \lambda[1] \models \varphi$;
$M, q \models \mathsf{E} \square \varphi$ iff there is a path λ such that $\lambda[0] = q$ and $M, \lambda[i] \models \varphi$ for every $i \geq 0$;
$M, q \models \mathsf{E} \varphi \mathcal{U} \psi$ iff there is a path λ with $\lambda[0] = q$, and position $i \geq 0$ such that $M, \lambda[i] \models \psi$ and $M, \lambda[j] \models \varphi$ for each $0 \leq j < i$.

Example 1 (Robots and Carriage). Consider the scenario depicted in Figure 1. Two robots push a carriage from opposite sides. As a result, the carriage can

---
[1] This variant of the language is sometimes called "vanilla" CTL. The broader language of CTL*, where no such restriction is imposed, is not discussed here.

move clockwise or anticlockwise, or it can remain in the same place – depending on who pushes with more force (and, perhaps, who refrains from pushing). To make our model of the domain discrete, we identify 3 different positions of the carriage, and associate them with states q_0, q_1, and q_2. The arrows in transition system M_0 indicate how the state of the system can change in a single step. We label the states with propositions $\mathsf{pos}_0, \mathsf{pos}_1, \mathsf{pos}_2$, respectively, to allow for referring to the current position of the carriage in the object language.

As an example CTL property, we have $M_0, q_0 \models \mathsf{E}\Diamond\mathsf{pos}_1$: in state q_0, there is a path such that the carriage will reach position 1 sometime in the future. Of course, the same is not true for *all* paths, so we also have that $M_0, q_0 \models \neg\mathsf{A}\Diamond\mathsf{pos}_1$.

2.2 ATL: A Logic of Strategic Ability

ATL [2,3] is a generalization of CTL, in which path quantifiers are replaced with so called *cooperation modalities*. Formula $\langle\!\langle A \rangle\!\rangle\varphi$ expresses that coalition A has a collective strategy to enforce φ. The recursive definition of ATL formulae is:

$$\varphi ::= p \mid \neg\varphi \mid \varphi \wedge \varphi \mid \langle\!\langle A \rangle\!\rangle \bigcirc \varphi \mid \langle\!\langle A \rangle\!\rangle \Box \varphi \mid \langle\!\langle A \rangle\!\rangle \varphi \mathcal{U} \varphi.$$

The semantics of ATL is defined in a variant of transition systems where transitions are labeled with combinations of actions, one per agent. Formally, a *concurrent game structure* (CGS) is a tuple $M = \langle \mathrm{Agt}, St, \Pi, \pi, Act, d, o \rangle$ which includes a nonempty finite set of all agents $\mathrm{Agt} = \{1, \ldots, k\}$, a nonempty set of states St, a set of atomic propositions Π and their valuation π, and a nonempty finite set of (atomic) actions Act. Function $d : \mathrm{Agt} \times St \to 2^{Act}$ defines nonempty sets of actions available to agents at each state, and o is a (deterministic) transition function that assigns the outcome state $q' = o(q, \alpha_1, \ldots, \alpha_k)$ to state q and a tuple of actions $\langle \alpha_1, \ldots, \alpha_k \rangle$, $\alpha_i \in d(i, q)$, that can be executed by Agt in q. So, it is assumed that all the agents execute their actions synchronously; the combination of the actions, together with the current state, determines the next transition of the system.

A *strategy* of agent a is a conditional plan that specifies what a is going to do in each possible state. Thus, a strategy can be represented with a function $s_a : St \to Act$, such that $s_a(q) \in d_a(q)$. A *collective strategy* for a group of agents $A = \{a_1, ..., a_r\}$ is simply a tuple of strategies $s_A = \langle s_{a_1}, ..., s_{a_r} \rangle$, one per agent from A.[2] By $s_A[a]$, we will denote agent a's part of the collective strategy s_A. Function $out(q, s_A)$ returns the set of all paths that may occur when agents A execute strategy s_A from state q onward:

$out(q, s_A) = \{\lambda = q_0 q_1 q_2... \mid q_0 = q$ and for each $i = 1, 2, ...$ there exists a tuple of agents' decisions $\langle \alpha_{a_1}^{i-1}, ..., \alpha_{a_k}^{i-1} \rangle$ such that $\alpha_a^{i-1} \in d_a(q_{i-1})$ for every $a \in \mathrm{Agt}$, and $\alpha_a^{i-1} \in s_A[a](q_{i-1})$ for every $a \in A$, and $o(q_{i-1}, \alpha_{a_1}^{i-1}, ..., \alpha_{a_k}^{i-1}) = q_i\}$.

---

[2] This is an important deviation from the original semantics of ATL [2,3], where strategies assign agents' choices to *sequences* of states. While the choice of one or another notion of strategy affects the semantics of most extensions of ATL (e.g. for abilities under imperfect information), it should be pointed out that both types of strategies yield equivalent semantics for "pure" ATL [17].

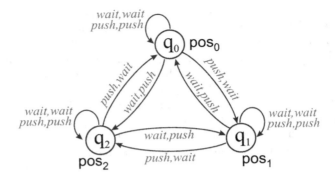

Fig. 2. The robots and the carriage: a concurrent game structure M_1

The semantics of cooperation modalities is defined through the clauses below. Informally speaking, $M, q \models \langle\!\langle A \rangle\!\rangle \Phi$ iff there exists a collective strategy s_A such that Φ holds for all computations from $out(q, s_A)$.

$M, q \models \langle\!\langle A \rangle\!\rangle \bigcirc \varphi$ iff there is a collective strategy s_A such that, for each path $\lambda \in out(s_A, q)$, we have $M, \lambda[1] \models \varphi$;

$M, q \models \langle\!\langle A \rangle\!\rangle \Box \varphi$ iff there exists s_A such that, for each $\lambda \in out(s_A, q)$, we have $M, \lambda[i] \models \varphi$ for every $i \geq 0$;

$M, q \models \langle\!\langle A \rangle\!\rangle \varphi \, \mathcal{U} \, \psi$ iff there exists s_A such that, for each $\lambda \in out(s_A, q)$, there is $i \geq 0$ for which $M, \lambda[i] \models \psi$, and $M, \lambda[j] \models \varphi$ for each $0 \leq j < i$.

Example 2 (Robots and Carriage, ctd.). Transition system M_0 enabled us to study the evolution of the system as a whole. However, it did not allow us to represent *who* can do *what*, and how the possible actions of the agents interact. Concurrent game structure M_1, presented in Figure 2, fills the gap. We assume that each robot can either push (action *push*) or refrain from pushing (action *wait*). Moreover, they both use the same force when pushing. Thus, if the robots push simultaneously or wait simultaneously, the carriage does not move. When only one of the robots is pushing, the carriage moves accordingly.

As the outcome of each robot's action depends on the current action of the other robot, no agent can make sure that the carriage moves to any particular position. So, we have for example that $M_1, q_0 \models \neg \langle\!\langle 1 \rangle\!\rangle \Diamond \mathsf{pos}_1$. On the other hand, the agent can at least make sure that the carriage will *avoid* particular positions. For instance, it holds that $M_1, q_0 \models \langle\!\langle 1 \rangle\!\rangle \Box \neg \mathsf{pos}_1$, the right strategy being $s_1(q_0) = wait, s_1(q_2) = push$ (the action that we specify for q_1 is irrelevant).

Note that the CTL path quantifiers A and E can be embedded in ATL in the following way: $\mathsf{A}\varphi \equiv \langle\!\langle \varnothing \rangle\!\rangle \varphi$ and $\mathsf{E}\varphi \equiv \langle\!\langle \mathbb{A}\mathrm{gt} \rangle\!\rangle \varphi$.

2.3   Strategic Abilities under Imperfect Information

ATL and its models include no way of addressing uncertainty that an agent or a process may have about the current situation. Here, we take Schobbens'

ATL$_{ir}$ [17] as the "core", minimal ATL-based language for strategic ability under imperfect information. ATL$_{ir}$ includes the same formulae as ATL, only the cooperation modalities are presented with a subscript: $\langle\!\langle A \rangle\!\rangle_{ir}$ to indicate that they address agents with imperfect information and imperfect recall. Models of ATL$_{ir}$, *imperfect information concurrent game structures* (i-CGS), can be seen as concurrent game structures augmented with a family of indistinguishability relations $\sim_a \subseteq St \times St$, one per agent $a \in$ Agt. The relations describe agents' uncertainty: $q \sim_a q'$ means that, while the system is in state q, agent a considers it possible that it is in q'. Each $\sim_a$ is assumed to be an equivalence. It is also required that agents have the same choices in indistinguishable states: if $q \sim_a q'$ then $d(a, q) = d(a, q')$.

Again, a *strategy* of an agent a is a conditional plan that specifies what a is going to do in each possible state. An executable (deterministic) plan must prescribe the same choices for indistinguishable states. Therefore ATL$_{ir}$ restricts the strategies that can be used by agents to the set of so called uniform strategies. A *uniform strategy* of agent a is defined as a function $s_a : St \to Act$, such that: (1) $s_a(q) \in d(a, q)$, and (2) if $q \sim_a q'$ then $s_a(q) = s_a(q')$. A collective strategy is uniform if it contains only uniform individual strategies. Again, function $out(q, s_A)$ returns the set of all paths that may result from agents A executing strategy s_A from state q onward. The semantics of cooperation modalities in ATL$_{ir}$ is defined as follows:

$M, q \models \langle\!\langle A \rangle\!\rangle_{ir} \bigcirc \varphi$ iff there exists a uniform collective strategy s_A such that, for each $a \in A$, q' such that $q \sim_a q'$, and path $\lambda \in out(s_A, q')$, we have $M, \lambda[1] \models \varphi$;

$M, q \models \langle\!\langle A \rangle\!\rangle_{ir} \Box \varphi$ iff there is a uniform s_A such that, for each $a \in A$, q' such that $q \sim_a q'$, and $\lambda \in out(s_A, q')$, we have $M, \lambda[i] \models \varphi$ for each $i \geq 0$;

$M, q \models \langle\!\langle A \rangle\!\rangle_{ir} \varphi \mathcal{U} \psi$ iff there exists a uniform strategy s_A such that, for each $a \in A$, q' such that $q \sim_a q'$, and $\lambda \in out(s_A, q')$, there is $i \geq 0$ for which $M, \lambda[i] \models \psi$, and $M, \lambda[j] \models \varphi$ for every $0 \leq j < i$.

That is, $\langle\!\langle A \rangle\!\rangle_{ir} \Phi$ if agents A have a uniform strategy such that, for each path that can possibly result from execution of the strategy *according to at least one agent from A*, Φ is the case.

Example 3 (Robots and Carriage, ctd.). We refine the scenario from Examples 1 and 2 by restricting perception of the robots. Namely, we assume that robot 1 is only able to observe the color of the surface on which it is standing, and robot 2 perceives only the texture (cf. Figure 3). In consequence, the first robot can distinguish between position 0 and position 1, but positions 0 and 2 look the same to it. Likewise, the second robot can distinguish between positions 0 and 2, but not 0 and 1. We also assume that the agents are memoryless: every time they come back to the same position, their knowledge of the current situation is limited in the same way as before.

With its observational capabilities restricted in such way, no agent can make the carriage reach or avoid any selected states singlehandedly. E.g., we have that $M_2, q_0 \models \neg\langle\!\langle 1 \rangle\!\rangle_{ir} \Box \neg \mathsf{pos}_1$; note in particular that strategy s_1 from Example 2

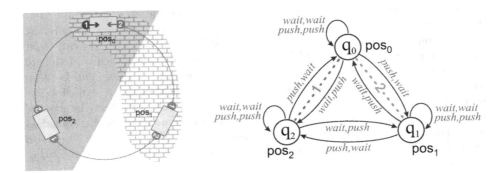

Fig. 3. Two robots and a carriage: a schematic view (left) and an imperfect information concurrent game structure M_2 that models the scenario (right)

cannot be used here because it is not uniform (indeed, the strategy tells robot 1 to wait in q_0 and push in q_2 but both states look the same to it). The robots cannot even be sure to achieve the task together: $M_2, q_0 \models \neg \langle\!\langle 1, 2 \rangle\!\rangle_{ir} \Box \mathsf{pos}_1$ (when in q_0, robot 2 considers it possible that the current state of the system is q_1, in which case all the hope is gone). So, do the robots know how to play to achieve anything? Yes, for example they know how to make the carriage *reach* a particular state eventually: $M_2, q_0 \models \langle\!\langle 1, 2 \rangle\!\rangle_{ir} \Diamond \mathsf{pos}_1$ etc. – it suffices that one of the robots pushes all the time and the other waits all the time.

3   A Survey of Model Checking Complexity Results

3.1   Model Checking Is Easy

It has been known for a long time that formulae of CTL can be checked in time linear with respect to the size of the model and the length of the formula. One of the main results concerning ATL states that its formulae can also be model-checked in deterministic linear time.

Theorem 1 ([5]). *Model checking* CTL *is* **P***-complete, and can be done in time* $\mathbf{O}(ml)$*, where m is the number of transitions in the model and l is the length of the formula.*

Theorem 2 ([3]). *Model checking* ATL *is* **P***-complete, and can be done in time* $\mathbf{O}(ml)$*, where m is the number of transitions in the model and l is the length of the formula.*

The ATL model checking algorithm from [3] is presented in Figure 4. The algorithm uses the well-known fixpoint characterizations of strategic-temporal modalities:

$$\langle\!\langle A \rangle\!\rangle \Box \varphi \leftrightarrow \varphi \wedge \langle\!\langle A \rangle\!\rangle \bigcirc \langle\!\langle A \rangle\!\rangle \Box \varphi$$
$$\langle\!\langle A \rangle\!\rangle \varphi_1 \, \mathcal{U} \, \varphi_2 \leftrightarrow \varphi_2 \vee \varphi_1 \wedge \langle\!\langle A \rangle\!\rangle \bigcirc \langle\!\langle A \rangle\!\rangle \varphi_1 \, \mathcal{U} \, \varphi_2$$

---

function $mcheck(M, \varphi)$.

Returns the set of states in model $M = \langle \text{Agt}, St, \Pi, \pi, o \rangle$ for which formula φ holds.

case $\varphi \in \Pi$:   return $\pi(p)$

case $\varphi = \neg\psi$:   return $St \setminus mcheck(M, \psi)$

case $\varphi = \psi_1 \vee \psi_2$:   return $mcheck(M, \psi_1) \cup mcheck(M, \psi_2)$

case $\varphi = \langle\!\langle A \rangle\!\rangle \bigcirc \psi$:   return $pre(M, A, mcheck(M, \psi))$

case $\varphi = \langle\!\langle A \rangle\!\rangle \Box \psi$:

  $Q_1 := St;$ $Q_2 := mcheck(M, \psi);$ $Q_3 := Q_2;$

  while $Q_1 \nsubseteq Q_2$

  do $Q_1 := Q_2;$ $Q_2 := pre(M, A, Q_1) \cap Q_3$ **od**;

  return Q_1

case $\varphi = \langle\!\langle A \rangle\!\rangle \psi_1 \, \mathcal{U} \, \psi_2$:

  $Q_1 := \varnothing;$ $Q_2 := mcheck(M, \psi_1);$

  $Q_3 := mcheck(M, \psi_2);$

  while $Q_3 \nsubseteq Q_1$

  do $Q_1 := Q_1 \cup Q_3;$ $Q_3 := pre(M, A, Q_1) \cap Q_2$ **od**;

  return Q_1

end case

---

function $pre(M, A, Q)$.

Auxiliary function; returns the exact set of states Q' such that, when the system is in a state $q \in Q'$, agents A can cooperate and enforce the next state to be in Q.

return $\{q \mid \exists \alpha_A \forall \alpha_{\text{Agt} \setminus A} \; o(q, \alpha_A, \alpha_{\text{Agt} \setminus A}) \in Q\}$

Fig. 4. The ATL model checking algorithm from [3]

and computes a winning strategy step by step (if it exists). That is, it starts with the appropriate candidate set of states ($\emptyset$ for $\mathcal{U}$ and St for $\Box$), and iterates over A's one-step abilities until the set gets stable. It is easy to see that the algorithm needs to traverse each transition at most once per subformula of φ.

In contrast, analogous fixpoint characterizations do not hold for ATL$_{ir}$ modalities because the choice of a particular action at state q has non-local consequences: it automatically fixes choices at all states q' indistinguishable from q for the coalition A. Moreover, agents' ability to *identify* a strategy as winning also varies throughout the game in an arbitrary way (agents can learn as well as forget). This suggests that winning strategies cannot be synthesized incrementally, which is indeed confirmed by the following (rather pessimistic) result.

Theorem 3 ([17,11]). *Model checking* ATL$_{ir}$ *is* Δ_2^P-*complete in the number of transitions in the model and the length of the formula.*

Still, model checking CTL and ATL appear to be tractable. So... let's model check! Unfortunately, it turns out to be not as easy as it seems.

3.2   Model Checking Is Harder

The results from [5,3] are certainly attractive, but it should be kept in mind that they are only relative to the size of models and formulae, and these can be very large for most application domains. Indeed, it is known that the number of states in a model is usually exponential in the size of a higher-level description of the problem domain for both CTL and ATL models. We will discuss this case in more detail in Section 3.3. In this section, we still consider model checking with respect to transition systems, concurrent game structures, and i-CGS's,[3] but we measure the size of the input in a slightly different way.

The size of a model has been defined as the number of transitions (m). Why not states then? For CTL this would not change the picture much. Let us denote the number of states by n; then, for any unlabeled transition system we have that $m = \mathbf{O}(n^2)$. In consequence, CTL model checking is in $\mathbf{P}$ also with respect to the number of states in the model (and the length of the formula). For ATL, however, the situation is different.

Observation 1 ([3,10]). *Let n be the number of states in a concurrent game structure M. The number of transitions in M is not bounded by n^2, because transitions are labeled with tuples of agents' choices.*

Let k denote the number of agents, and d the maximal number of available decisions per agent per state. Then, $m = \mathbf{O}(nd^k)$. In consequence, the ATL model checking algorithm from [3] runs in time $\mathbf{O}(nd^k l)$, and hence its complexity is exponential if the number of agents is a parameter of the problem.

As we see, the complexity of $\mathbf{O}(ml)$ may (but does not have to) include potential intractability *even on the level of explicit models* if the size of models is defined in terms of states rather than transitions, and the number of agents is a parameter of the problem.

Corollary 1 (of Theorem 1). CTL *model checking is $\mathbf{P}$-complete, and can be done in time $\mathbf{O}(n^2 l)$.*

Theorem 4 ([10,14]). *Model checking ATL is $\mathbf{\Delta_3^P}$-complete with respect to the number of states and agents, and the length of the formula.*

It also turns out that model checking of abilities under imperfect information looks no harder than perfect information from this perspective.

Theorem 5 ([11]). *Model checking ATL$_{ir}$ is $\mathbf{\Delta_3^P}$-complete with respect to the number of states and agents, and the length of the formula.*

3.3   Model Checking Is Hard

Sections 3.1 and 3.2 presented complexity results for model checking CTL, ATL, and ATL$_{ir}$ with respect to explicit models. Most multi-agent systems, however, are characterized by an immensely huge state space and transition relation. In such cases, one would like to define the model in terms of a compact higher-level representation, plus an unfolding procedure that defines the relationship between

---

[3]   Such structures are sometimes called *explicit models* [15] because global states and global transitions are represented explicitly in them.

representations and actual models of the logic (and hence also the semantics of the logic with respect to the compact representations). Of course, unfolding a higher-level description to an explicit model involves usually an exponential blowup in its size.

Consider, for example, a system whose state space is defined by r Boolean variables (binary attributes). Obviously, the number of global states in the system is $n = 2^r$. A more general approach is presented in [12], where the "higher level description" is defined in terms of so called *concurrent programs*, that can be used for simulating Boolean variables, but also processes or agents acting in parallel. A concurrent program P is composed of k concurrent processes, each described by a labeled transition system $P_i = \langle St_i, Act_i, \mathcal{R}_i, \Pi_i, \pi_i \rangle$, where St_i is the set of local states of process i, Act_i is the set of local actions, $\mathcal{R}_i \subseteq St_i \times Act_i \times St_i$ is a transition relation, and Π_i, π_i are the set of local propositions and their valuation. The behavior of program P is given by the product automaton of $P_1, ..., P_k$ under the assumption that processes work asynchronously, actions are interleaved, and synchronization is obtained through common action names.

Theorem 6 ([12]). *Model checking* CTL *in concurrent programs is* **PSPACE**-*complete with respect to the number of local states and agents (processes), and the length of the formula.*

Concurrent programs seem sufficient to reason about purely temporal properties of systems, but not quite so for reasoning about agents' strategies and abilities. For the latter kind of analysis, we need to allow for more sophisticated interference between agents' actions (and enable modeling agents that play synchronously). Here, we use *modular interpreted systems* [9,8], that draw inspiration from interpreted systems [7], reactive modules [1], and are in many respects similar to ISPL specifications [16]. A modular interpreted system (MIS) is defined as a tuple $s = \langle \text{Agt}, env, Act, \mathcal{In} \rangle$, where $\text{Agt} = \{a_1, ..., a_k\}$ is a set of agents, env is the environment, Act is a set of actions, and $\mathcal{In}$ is a set of symbols called *interaction alphabet*. Each agent has the following internal structure: $a_i = \langle St_i, d_i, out_i, in_i, o_i, \Pi_i, \pi_i \rangle$, where:

- St_i is a set of local states,
- $d_i : St_i \rightarrow 2^{Act}$ defines local availability of actions; for convenience of the notation, we additionally define the set of *situated actions* as $D_i = \{\langle q_i, \alpha \rangle \mid q_i \in St_i, \alpha \in d_i(q_i)\}$,
- out_i, in_i are *interaction functions*; $out_i : D_i \rightarrow \mathcal{In}$ refers to the influence that a given situated action (of agent a_i) may possibly have on the external world, and $in_i : St_i \times \mathcal{In}^k \rightarrow \mathcal{In}$ translates external manifestations of the other agents (and the environment) into the "impression" that they make on a_i's transition function depending on the local state of a_i,
- $o_i : D_i \times \mathcal{In} \rightarrow St_i$ is a (deterministic) local transition function,
- Π_i is a set of local propositions of agent a_i where we require that Π_i and Π_j are disjunct when $i \neq j$, and
- $\pi_i : \Pi_i \rightarrow 2^{St_i}$ is a valuation of these propositions.

The environment *env* has the same structure as an agent except that it does not perform actions.

The unfolding of a MIS s to a concurrent game structure follows by the synchronous product of the agents (and the environment) in s, with interaction symbols being passed between local transition functions at every step. The unfolding can also determine indistinguishability relations as $\langle q_1, ..., q_k, q_{env} \rangle \sim_i \langle q'_1, ..., q'_k, q'_{env} \rangle$ iff $q_i = q'_i$, thus yielding a full *i*CGS. This way semantics of both ATL and ATL$_{ir}$ is extended to MIS. Regarding model checking complexity, the following holds.

Theorem 7 ([18,8]). *Model checking* ATL *in modular interpreted systems is* **EXPTIME**-*complete with respect to the number of local states and agents, and the length of the formula.*

Theorem 8 ([9,8]). *Model checking* ATL$_{ir}$ *in modular interpreted systems is* **PSPACE**-*complete with respect to the number of local states and agents, and the length of the formula.*

3.4   Summary of the Results

A summary of complexity results for model checking temporal and strategic logics is given in the table below. Symbols n, m stand for the number of states and transitions in an explicit model; k is the number of agents in the model; l is the length of the formula, and n_{local} is the number of local states in a concurrent program or a modular interpreted system.

	m, l	n, k, l	n_{local}, k, l
CTL	**P**-complete [5]	**P**-complete [5]	**PSPACE**-complete [12]
ATL	**P**-complete [3]	$\mathbf{\Delta_3^P}$-compl. [10,14]	**EXPTIME**-compl. [18,8]
ATL$_{ir}$	$\mathbf{\Delta_2^P}$-compl. [17,11]	$\mathbf{\Delta_3^P}$-complete [11]	**PSPACE**-complete [9,8]

Note that the results for ATL and ATL$_{ir}$ form an intriguing pattern. When we compare model checking agents with perfect vs. imperfect information, the first problem appears to be much easier against explicit models measured with the number of transitions; next, we get the same complexity class against explicit models measured with the number of states and agents; finally, model checking imperfect information turns out to be *easier* than model checking perfect information for modular interpreted systems. Why can it be so?

The amount of available strategies (relative to the size of input parameters) is the crucial factor here. The number of all strategies is exponential in the number of global states; for uniform strategies, there are usually much less of them but still exponentially many in general. Thus, the fact that perfect information strategies can be synthesized incrementally has a substantial impact on the complexity of the problem. However, measured in terms of local states and agents, the number of all strategies is *doubly exponential*, while there are "only" exponentially many uniform strategies – which settles the results in favor of imperfect information. It must be also noted that *representation* of a concurrent game structure by a MIS can be in general more compact than that of an *i*CGS. In the latter case, the MIS is assumed to encode the epistemic relations explicitly. In the case of CGS, the epistemic aspect is ignored, which gives some extra room for encoding the transition relation more efficiently.

4 Conclusions

This paper recalls some important complexity results for model checking temporal and strategic properties of multi-agent systems. But, most of all, it tells a story with a moral. A single complexity result is often not enough to understand the *real* difficulty of the decision problem in question. When the perspective changes, so does the complexity class in which the problem belongs, and sometimes even its relative complexity with respect to other problems. Does it mean that theoretical complexity studies are worthless? Of course not, but the computational difficulty of a problem is usually more intricate than most computer scientists suspect.

There is yet another moral, too. Experimental studies where performance of algorithms and tools is measured in practice are no less needed than theoretical analysis.

References

1. Alur, R., Henzinger, T.A.: Reactive modules. Formal Methods in System Design 15(1), 7–48 (1999)
2. Alur, R., Henzinger, T.A., Kupferman, O.: Alternating-time Temporal Logic. In: Proceedings of the 38th Annual Symposium on Foundations of Computer Science (FOCS), pp. 100–109. IEEE Computer Society Press, Los Alamitos (1997)
3. Alur, R., Henzinger, T.A., Kupferman, O.: Alternating-time Temporal Logic. Journal of the ACM 49, 672–713 (2002)
4. Clarke, E.M., Emerson, E.A.: Design and synthesis of synchronization skeletons using branching time temporal logic. In: Proceedings of Logics of Programs Workshop. LNCS, vol. 131, pp. 52–71. Springer, Heidelberg (1981)
5. Clarke, E.M., Emerson, E.A., Sistla, A.P.: Automatic verification of finite-state concurrent systems using temporal logic specifications. ACM Transactions on Programming Languages and Systems 8(2), 244–263 (1986)
6. Emerson, E.A., Halpern, J.Y.: "sometimes" and "not never" revisited: On branching versus linear time temporal logic. Journal of the ACM 33(1), 151–178 (1986)
7. Fagin, R., Halpern, J.Y., Moses, Y., Vardi, M.Y.: Reasoning about Knowledge. MIT Press, Cambridge (1995)
8. Jamroga, W., Ågotnes, T.: Modular interpreted systems: A preliminary report. Technical Report IfI-06-15, Clausthal University of Technology (2006)
9. Jamroga, W., Ågotnes, T.: Modular interpreted systems. In: Proceedings of AAMAS 2007, pp. 892–899 (2007)
10. Jamroga, W., Dix, J.: Do agents make model checking explode (computationally)? In: Pěchouček, M., Petta, P., Varga, L.Z. (eds.) CEEMAS 2005. LNCS, vol. 3690, pp. 398–407. Springer, Heidelberg (2005)

11. Jamroga, W., Dix, J.: Model checking abilities of agents: A closer look. Theory of Computing Systems 42(3), 366–410 (2008)
12. Kupferman, O., Vardi, M.Y., Wolper, P.: An automata-theoretic approach to branching-time model checking. Journal of the ACM 47(2), 312–360 (2000)
13. Laroussinie, F.: About the expressive power of CTL combinators. Information Processing Letters 54(6), 343–345 (1995)
14. Laroussinie, F., Markey, N., Oreiby, G.: Expressiveness and complexity of ATL. Technical Report LSV-06-03, CNRS & ENS Cachan, France (2006)
15. McMillan, K.L.: Symbolic Model Checking: An Approach to the State Explosion Problem. Kluwer Academic Publishers, Dordrecht (1993)
16. Raimondi, F.: Model Checking Multi-Agent Systems. PhD thesis, University College London (2006)
17. Schobbens, P.Y.: Alternating-time logic with imperfect recall. Electronic Notes in Theoretical Computer Science 85(2) (2004)
18. van der Hoek, W., Lomuscio, A., Wooldridge, M.: On the complexity of practical ATL model checking. In: Stone, P., Weiss, G. (eds.) Proceedings of AAMAS 2006, pp. 201–208 (2006)

Programming Multi-agent Systems
(Extended Abstract)

Mehdi Dastani

Utrecht University
The Netherlands
mehdi@cs.uu.nl

Multi-agent systems consist of a number of interacting autonomous agents, each of which is capable of sensing its environment (including other agents) and deciding to act in order to achieve its own objectives. In order to guarantee the overall design objectives of multi-agent systems, the behavior of individual agents and their interactions need to be regulated and coordinated [23,29,30]. The development of multi-agent systems therefore requires programming languages that facilitate the implementation of individual agents as well as mechanisms that control and regulate individual agents' behaviors. It also requires computational tools to test and verify programs that implement multi-agent systems [7].

In the last two decades, various multi-agent programming languages have been proposed to facilitate the implementation of multi-agent systems [6,7,11]. These programming languages differ in detail, despite their apparent similarities [8,9,15,16,18,20,21,22,24,25,28]. They differ from each other as they provide programming constructs for different, sometimes overlapping, sets of agent concepts and abstractions. The expressivity of the programming constructs for overlapping agent concepts and abstractions may differ from one programming language to another. This is partly due to the fact that they are based on different logics or use different technologies. Moreover, some of these programming languages have formal semantics, but others provide only an informal explanation of the intended meaning of their programming constructs. Also, some programming languages capture in their semantics specific rationality principles that underlie agent concepts (e.g., relating beliefs and goals), while such principles are assumed to be implemented explicitly by agent programmers in other programming languages. Finally, some agent-oriented programming languages are based on declarative style programming, some are based on imperative style programming, and yet others combine these programming styles.

An example of a multi-agent programming language which combines declarative and imperative programming styles is 2APL (A Practical Agent Programming Language)[11]. 2APL is a BDI-based agent-oriented programming language that is developed to facilitate the implementation of individual agents in terms of cognitive concepts such as beliefs, goals, plans, events, and reasoning rules. The interpreter of 2APL is basically a decision process that selects and performs actions based on a sense, reason, and act cycle, often called deliberation cycle. In order to verify 2APL programs and to check their correctness a logic

M. Fisher, F. Sadri, and M. Thielscher (Eds.): CLIMA IX, LNAI 5405, pp. 13–16, 2009.
© Springer-Verlag Berlin Heidelberg 2009

is developed that can be used to reason about such programs [1,2]. The logic, which is based on PDL (Propositional Dynamic Logic), can be used to reason about possible executions of agent programs. Also, a prototype implementation of the 2APL interpreter in Maude [10] (a rewriting logic software) is developed [4]. This implementation allows the use of the model checking tools that come with Maude and allows the verification of 2APL programs.

One of the challenges in the design and development of multi-agent systems is to control and coordinate the behavior and interaction of individual agent programs in order to guarantee the overall system design objectives. There are two general approaches to develop coordination mechanisms for multi-agent systems: endogenous and exogenous coordination approaches. In the endogenous coordination approach, the coordination mechanism is incorporated in the individual agent programs while in the exogenous coordination approach the coordination mechanism is a separate program/module ensuring a clear separation between individual agent programs and their coordination mechanisms. An advantage of the exogenous approach is that the overall system design objectives can be developed and verified independent of the individual agent programs. Exogenous coordination mechanisms for multi-agent systems can be designed and developed in terms of concepts such as action synchronization and resource access relation [3,23], but also in terms of social and organizational concepts (e.g., norms, roles, groups, responsibility) and processes (e.g., monitoring actions and sanctioning mechanisms) [14,17,19,26,27].

A specific proposal for the design and development of exogenous coordination mechanisms for multi-agent systems is the introduction of an organization-based programming language [12,13]. This proposed programming language is designed to facilitate the implementation of organization based coordination artifacts in terms of norms and sanctions. Such artifacts, also called multi-agent organizations, refer to norms as a way to signal when violations take place and sanctions as a way to respond in the case of violations. Basically, a norm-based artifact observes the actions performed by the individual agents, determines the violations caused by performing the actions, and possibly imposes sanctions. The operational semantics of this programming language make it easy to prototype it in Maude [10]. The Maude implementation of the interpreter of this organization-based coordination language [5] allows model checking organization-based programs. Finally, in order to reason about organization-based coordination programs, a logic is devised to reason about such programs [12].

References

1. Alechina, N., Dastani, M., Logan, B., Meyer, J.-J.: A logic of agent programs. In: Proceedings of the twenty-second conference on Artificial Intelligence (AAAI 2007). AAAI Press, Menlo Park (2007)
2. Alechina, N., Dastani, M., Logan, B., Meyer, J.-J.: Reasoning about agent deliberation. In: Proceedings of eleventh international conference on principles of knowledge representation and Reasoning (KR 2008) (2008)

3. Arbab, F.: Reo: a channel-based coordination model for component composition. Mathematical Structures in Computer Science 14(3), 329–366 (2004)
4. Arbab, F., Astefanoaei, L., de Boer, F.S., Dastani, M., Meyer, J.-J., Tinnermeier, N.: Reo connectors as coordination artifacts in 2apl systems. In: Bui, T.D., Ho, T.V., Ha, Q.T. (eds.) PRIMA 2008. LNCS, vol. 5357, pp. 42–53. Springer, Heidelberg (2008)
5. Astefanoaei, L., Dastani, M., Meyer, J.-J., de Boer, F.S.: A verification framework for normative multi-agent systems. In: Bui, T.D., Ho, T.V., Ha, Q.T. (eds.) PRIMA 2008. LNCS, vol. 5357, pp. 54–65. Springer, Heidelberg (2008)
6. Bordini, R., Braubach, L., Dastani, M., Seghrouchni, A.E.F., Gomez-Sanz, J., Leite, J., O'Hare, G., Pokahr, A., Ricci, A.: A survey of programming languages and platforms for multi-agent systems. Informatica 30, 33–44 (2006)
7. Bordini, R.H., Dastani, M., Dix, J., El Fallah Seghrouchni, A. (eds.): Multi-Agent Programming: Languages, Platforms and Applications. Springer, Berlin (2005)
8. Bordini, R.H., Wooldridge, M., Hübner, J.F.: Programming Multi-Agent Systems in AgentSpeak using Jason (Wiley Series in Agent Technology). John Wiley & Sons, Chichester (2007)
9. Bracciali, A., Demetriou, N., Endriss, U., Kakas, A., Lu, W., Sadri, P.M.F., Stathis, K., Terreni, G., Toni, F.: The KGP model of agency for global computing: Computational model and prototype implementation. In: Priami, C., Quaglia, P. (eds.) GC 2004. LNCS, vol. 3267, pp. 340–367. Springer, Heidelberg (2005)
10. Clavel, M., Durán, F., Eker, S., Lincoln, P., Martí-Oliet, N., Meseguer, J., Quesada, J.F.: Maude: Specification and programming in rewriting logic. Theoretical Computer Science (2001)
11. Dastani, M.: 2APL: a practical agent programming language. International Journal of Autonomous Agents and Multi-Agent Systems (JAAMAS) 16(3), 214–248 (2008)
12. Dastani, M., Grossi, D., Tinnemeier, N., Meyer, J.-J.: A programming language for normative multi-agent systems. In: Proceedings of the Workshop on Knowledge Representation for Agents and Multi-Agent Systems (KRAMAS 2008) (2008)
13. Dastani, M., Tinnemeier, N., Meyer, J.-J.: A programming language for normative multi-agent systems. In: Dignum, V. (ed.) Multi-Agent Systems: Semantics and Dynamics of Organizational Models, IGI Global (to be published) (2009)
14. Esteva, M., Rodríguez-Aguilar, J., Rosell, B., Arcos, J.: Ameli: An agent-based middleware for electronic institutions. In: Proc. of AAMAS 2004, New York, US (2004)
15. Fisher, M.: METATEM: The story so far. In: Bordini, R.H., Dastani, M., Dix, J., El Fallah Seghrouchni, A. (eds.) PROMAS 2005. LNCS, vol. 3862, pp. 3–22. Springer, Heidelberg (2006)
16. Giacomo, G.D., Lesperance, Y., Levesque, H.J.: Congolog, a concurrent programming language based on the situation calculus. Artificial Intelligence 121(1-2), 109–169 (2000)
17. Grossi, D., Dignum, F., Dastani, M., Royakkers, L.: Foundations of organizational structures in multiagent systems. In: Proc. of AAMAS 2005, pp. 690–697. ACM Press, New York (2005)
18. Hindriks, K.V., Boer, F.S.D., der Hoek, W.V., Meyer, J.-J.C.: Agent programming in 3APL. Autonomous Agents and Multi-Agent Systems 2(4), 357–401 (1999)
19. Hübner, J.F., Sichman, J.S., Boissier, O.: Moise+: Towards a structural functional and deontic model for mas organization. In: Proc. of AAMAS 2002, Bologna, Italy. ACM Press, Bologna (2002)

20. Kakas, A., Mancarella, P., Sadri, F., Stathis, K., Toni, F.: The KGP model of agency. In: The 16th European Conference on Artificial Intelligence, pp. 33–37 (2004)
21. Leite, J.A., Alferes, J.J., Pereira, L.M.: Minerva — A dynamic logic programming agent architecture. In: Meyer, J.-J.C., Tambe, M. (eds.) ATAL 2001. LNCS, vol. 2333, p. 141. Springer, Heidelberg (2002)
22. Pokahr, A., Braubach, L., Lamersdorf, W.: Jadex: A BDI reasoning engine. In: Multi-Agent Programming: Languages, Platforms and Applications. Kluwer Academic Publishers, Dordrecht (2005)
23. Ricci, A., Viroli, M., Omicini, A.: Give agents their artifacts: the A&A approach for engineering working environments in MAS. In: Proc. of AAMAS 2007, pp. 1–3. ACM, New York (2007)
24. Sadri, F.: Using the KGP model of agency to design applications. In: Toni, F., Torroni, P. (eds.) CLIMA 2005. LNCS, vol. 3900, pp. 165–185. Springer, Heidelberg (2006)
25. Sardina, S., Giacomo, G.D., Lespérance, Y., Levesque, H.J.: On the semantics of deliberation in indigolog from theory to implementation. Annals of Mathematics and Artificial Intelligence 41(2-4), 259–299 (2004)
26. Sierra, C., Rodríguez-Aguilar, J., Noriega, P., Esteva, M., Arcos, J.L.: Engineering multi-agent systems as electronic institutions. UPGrade 4 (2004)
27. Tinnemeier, N.A., Dastani, M., Meyer, J.-J.: Orwell's nightmare for agents? programming multi-agent organisations. In: Proceedings of the workshop on Programming Multi-Agent Systems (ProMAS 2008). Springer, Heidelberg (2008)
28. Winikoff, M.: JACK^{TM} intelligent agents: An industrial strength platform. In: Multi-Agent Programming: Languages, Platforms and Applications. Kluwer, Dordrecht (2005)
29. Woolridge, M.: Introduction to Multiagent Systems. John Wiley & Sons, Inc., Chichester (2002)
30. Zambonelli, F., Jennings, N., Wooldridge, M.: Organisational rules as an abstraction for the analysis and design of multi-agent systems. IJSEKE 11(3), 303–328 (2001)

Verifying Agents' Conformance with Multiparty Protocols

Laura Giordano[1] and Alberto Martelli[2]

[1] Dipartimento di Informatica, Università del Piemonte Orientale, Alessandria
[2] Dipartimento di Informatica, Università di Torino, Torino

Abstract. The paper deals with the problem of agents conformance with multiparty protocols. We introduce a notion of conformance of a set of k agents with a multiparty protocol with k roles, which requires the agents to be interoperable and to produce correct executions of the protocol. We introduce conditions that enable each agent to be independently verified with respect to the protocol. We assume that protocols are specified in a temporal action theory and we show that the problem of verifying the conformance of an agent with a protocol can be solved by making use of automata based techniques. Protocols with nonterminating computations, modeling reactive agents, can also be captured in this framework.

1 Introduction

In an open environment, the interaction of agents is ruled by interaction protocols on which agents commonly agree. An important issue, in this regard, concerns agent conformance with the protocol. Although agent policy may somehow deviate from the behavior dictated by the protocol, in some cases we want, nevertheless, to regard the policy as being compatible with the protocol. A related issue concerns the *interoperability* of agents in an open environment. The need for conditions to guarantee that a set of agents may interact properly, has led to the introduction of different notions of *compatibility* among agents [4] as well as to the definition of notions of *conformance* of an agent with a protocol [2,5,11,17].

In this paper, we define a notion of agent conformance for the general case of multiparty protocols. This notion must assure that a set of agents, that are conformant with a protocol, interoperate (in particular, they do not get stuck) and that their interactions produce correct executions of the protocol.

In our proposal, the specification of agents and protocols is given in a temporal action theory [9,13], by means of temporal constraints, and the communication among agents is synchronous. Protocols are given a *declarative specification* consisting of: (i) the specification of communicative actions by means of their effects and preconditions on the social state which, in particular, includes commitments; (ii) a set of temporal constraints, which specify the wanted interactions (and, under this respect, our approach to protocol specification is similar to the

M. Fisher, F. Sadri, and M. Thielscher (Eds.): CLIMA IX, LNAI 5405, pp. 17–36, 2009.

one proposed in *DecSerFlow* [21]). Protocols with nonterminating computations, modeling reactive services [8], can also be captured in this framework.

We define a multiparty protocol P with k roles, by separately specifying the behavior of all roles $P_1, \ldots, P_k$ in the protocol. We then introduce a notion of *interoperability* among a set of agents, which guarantees the agents to interact properly. More precisely, each agent can freely choose among its possible emissions without the computation getting stuck.

Given a multiparty protocol P, we define a notion of *conformance of a set of agents* $A_1, \ldots, A_k$ with P: agents $A_1, \ldots, A_k$ interoperate and their interaction only produces runs of the protocol P. Verifying the conformance of a set of agents all together, however, is not feasible in an open environment, as, in general, the internal behavior of all agents participating in a protocol is not known. The verification of each agent participating in the protocol must be done independently.

In this paper, we introduce a definition of *conformance of a single agent* A_i *(playing role i) with the protocol* P. We will see that verifying an agent A_i with respect with its role P_i is not sufficient to guarantee the interoperability of a set of conformant agents in the multiparty case, unless a rather narrow notion of conformance is adopted. Indeed, our notion of conformance of an agent A_i with a protocol P is defined by comparing the runs of agent A_i, when executed in the context of the protocol, and the runs of P itself. We prove that a set of agents which are independently conformant with the protocol P are guaranteed to be interoperable and to produce correct executions of P.

Starting from a specification of the protocol in a temporal logic, the problem of verifying the conformance of an agent with a protocol can be solved by making use of an automata based approach. In particular, interoperability can be checked by working on the Büchi automaton which can be extracted from the logical specification of the protocol.

2   Protocol Specification

The specification of interaction protocols we adopt is based on Dynamic Linear Time Temporal Logic (DLTL) [15], a linear time temporal logic which extends LTL by allowing the until operator to be indexed by programs in Propositional Dynamic Logic (PDL). DLTL allows until formulas of the form $\alpha \mathcal{U}^\pi \beta$, where the program $\pi \in Prg(\Sigma)$ is a regular expression built from a set Σ of atomic actions. More precisely, $Prg(\Sigma) ::= a \mid \pi_1 + \pi_2 \mid \pi_1; \pi_2 \mid \pi^*$, where $a \in \Sigma$ and π_1, π_2, π range over $Prg(\Sigma)$.

As for LTL, DLTL models are infinite linear sequences of worlds (propositional interpretations), each one reachable from the initial world by a finite sequence τ of actions in the alphabet Σ. More precisely, a model $M = (\sigma, V)$ consists of an infinite sequence of actions σ over Σ (the sequence of actions executed from the initial world) and a valuation function V, defining the interpretation of propositions at each world τ (where τ is a prefix of σ).

In the following, we denote by $prf(\sigma)$ the set of all finite prefixes of σ (the worlds) and, for each regular program π, we denote by $[[\pi]]$ the set of finite sequences associated with π. Given a model $M = (\sigma, V)$, a finite word $\tau \in prf(\sigma)$ and a formula α, the *satisfiability of a formula α at τ in M*, written $M, \tau \models \alpha$, is defined as usual for the classical connectives. Moreover:

- $M, \tau \models p$ iff $p \in V(\tau)$;
- $M, \tau \models \alpha \mathcal{U}^\pi \beta$ iff there exists $\tau' \in [[\pi]]$ such that $\tau\tau' \in prf(\sigma)$ and $M, \tau\tau' \models \beta$. Moreover, for every τ'' such that $\varepsilon \leq \tau'' < \tau'$, $M, \tau\tau'' \models \alpha$.

A formula $\alpha \mathcal{U}^\pi \beta$ is true at a world τ if "α until β" is true on a finite stretch of behavior which is in the linear time behavior of the program π.

The derived modalities $\langle \pi \rangle$ and $[\pi]$ can be defined as follows: $\langle \pi \rangle \alpha \equiv \top \mathcal{U}^\pi \alpha$ and $[\pi]\alpha \equiv \neg \langle \pi \rangle \neg \alpha$. When π is Σ^* (representing all finite actions sequences), we replace $\langle \pi \rangle$ with $\Diamond$ and $[\pi]$ with $\Box$. Furthermore, the $\bigcirc$ (next) operator can be defined as $\bigcirc \alpha \equiv \bigvee_{a \in \Sigma} \langle a \rangle \alpha$. As shown in [15], DLTL(Σ) is strictly more expressive than LTL(Σ). The satisfiability and validity problems for DLTL are PSPACE complete problems [15].

In this paper, we make use of the Product version of DLTL, $DLTL^\otimes$ [14], which allows to describe the behavior of a network of sequential agents which coordinate their activities by performing common actions together. There are k agents $1, \ldots, k$, and a *distributed alphabet* $\tilde{\Sigma} = \{\Sigma_i\}_{i=1}^k$, a family of (possibly non-disjoint) alphabets, with each Σ_i a non-empty, finite set of actions (Σ_i is the set of actions which require the participation of agent i).

Atomic propositions are introduced in a local fashion, by introducing a non-empty set of atomic propositions $\mathcal{P}$. For each atomic proposition $p \in \mathcal{P}$ and agent i, p_i represents the "local" view of the proposition p at i, and is evaluated in the local state of agent i. The formulas of the language are obtained as the boolean combination of the formulas of $DLTL_i^\otimes$ which can be constructed on the alphabet (actions and propositions) of each agent i, using the modalities $\mathcal{U}_i^\pi$, $\langle \pi \rangle_i$, $[\pi]_i$, $\bigcirc_i$, $\Diamond_i$ and $\Box_i$. A $DLTL^\otimes$ model is a pair (σ, V), where $V = \{V_i\}_{i=1}^k$ is a family of valuation functions, one for each agent i. The satisfiability of the formulas of $DLTL_i^\otimes$ is evaluated by making use of V_i and of the projection $\sigma|_i$ of σ to Σ_i (where $\sigma|_i$ is the sequence obtained by erasing from σ all occurrences of symbols that are not in Σ_i).

We illustrate how a protocol can be specified in this framework trough the specification of a *Purchase protocol*.

Example 1. We have three roles: the *merchant (mr)*, the *customer (ct)* and the *bank (bk)*. *ct* sends a request to *mr*; *mr* replies with an offer or by saying that the requested good is not available. If *ct* receives the offer, it may either accept the offer and send a payment request to *bk*, or refuse the offer. If *ct* accepts the offer, then *mr* delivers the goods. If *ct* requires *bk* to pay *mr*, *bk* sends the payment. *ct* can send the request for payment to *bk* even before it has received the goods.

In this example, all actions are communicative actions: *sendRequest, sendOffer, sendNotAvail, sendAccept, sendRefuse, sendPaymentReqest, sendPayment*, each one belonging to the action alphabet of the sender and of the receiver. For instance, action *sendRequest* is both in Σ_{ct} (as ct is the sender of the request) and in Σ_{mr} (as mr is the receiver of the request). Communication is synchronous: roles communicate by synchronizing on the execution of communicative actions.

The Purchase protocol Pu is given by specifying separately the protocols of the three participating roles: P_{ct}, P_{mr} and P_{bk}. The role P_i in the protoocol is specified by a *domain description* D_i, which is a pair (Π_i, C_i), where Π_i is a set of formulas describing the effects and preconditions of the actions (the action theory) of role i, and C_i is a set of *constraints* that the executions of role i must satisfy. The approach is a generalization of the one proposed in [13].

Let us define, for instance, the *domain description* $D_{mr} = (\Pi_{mr}, C_{mr})$ of the merchant. We adopt a social approach where an interaction protocol is specified by describing the effects of communicative actions on the social state. The social state contains the domain specific fluents describing observable facts concerning the execution of the protocol (*request*, the customer has requested a quote, *accepted*, the customer has accepted the quote, etc.), but also special fluents to model *commitments* (and conditional commitments) among the roles [22,20]: $C(i, j, \alpha)$ says that role i is committed to role j to bring about α. Furthermore, a *conditional commitments* $CC(i, j, \beta, \alpha)$ says that role i is committed to role j to bring about α, if the condition β is brought about.

The action theory Π_{mr} consists of *action laws, causal laws, precondition laws*, and an *initial state*. Action laws $\mathcal{AL}_{mr}$ describe the effects of the execution of actions on the state. The action laws:

$$\Box_{mr}([sendOffer]_{mr}(offer \wedge CC(mr, ct, accept, goods)))$$
$$\Box_{mr}(request \rightarrow [sendOffer]_{mr}\neg request)$$

say that: when mr sends the quote for the good, then it commits to send the goods if ct accepts the request; and when mr sends the quote for the good, if there is a request, the request is cancelled.

Causal laws $\mathcal{CL}_{mr}$ are intended to expresses "causal" dependencies (or ramifications) among fluents. In this framework they are used to rule the dynamics of commitments. For instance, the *causal law*:

$$\Box_{mr}(\bigcirc_{mr}\alpha \rightarrow \bigcirc_{mr}\neg C(i, j, \alpha))$$

says that a commitment to bring about α is cancelled when α holds. Other causal laws are needed for dealing with conditional commitments.

Precondition laws $\mathcal{PL}$ have the form: $\Box(\alpha \rightarrow [a]\bot)$, meaning that the execution of an action a is not possible if α holds. The precondition law

$$\Box_{mr}(\neg request \rightarrow [sendOffer]_{mr}\bot)$$

says that an offer cannot be sent if a request has not been issued.

The *initial state* $\mathcal{IS}$ of the protocol defines the initial value of all the fluents. Here, we assume that the initial state is complete.

Action laws and causal laws describe the changes to the state. All other fluents which are not changed by action execution are assumed to persist unaltered to the next state. To cope with the *frame problem* [18] we use a completion construction $Comp$, which is applied given a domain description, introduces frame axioms in the style of the successor state axioms proposed by Reiter [19]. The completion construction $Comp$ is only applied to the action laws and to the causal laws [13]. Thus Π_{mr} is defined as $Comp(\mathcal{AL} \wedge \mathcal{CL}) \wedge \mathcal{PL} \wedge \mathcal{IS}$.

The second component $\mathcal{C}_{mr}$ of the domain description P_{mr} defines constraints as arbitrary temporal formulas of DLTL$_{mr}$. For instance, to model the fact that the merchant cannot send more than one offer, we introduce the constraint: $\neg \Diamond_{mr} \langle sendOffer \rangle_{mr} \Diamond_{mr} \langle sendOffer \rangle_{mr} \top$.

We are interested in those execution of the purchase protocol in which all commitments of all the roles have been fulfilled. In particular, for each commitment $C(i, j, \alpha)$ of which the merchant is a debtor or a creditor (i.e., $i = mr$ or $j = mr$), we add in $\mathcal{C}_{mr}$ the constraint: $\Box_{mr}(C(i, j, \alpha) \rightarrow \Diamond_{mr}(\alpha \vee \neg C(i, j, \alpha)))$, meaning that a commitment has to be fulfilled unless it is cancelled.

Given $D_{mr} = (\Pi_{mr}, \mathcal{C}_{mr})$ as defined above, we let $P_{mr} = \Pi_{mr} \wedge \mathcal{C}_{mr}$. Once the protocols P_{ct}, P_{mr} and P_{bk} have been defined, the specification Pu of the Purchase protocol can be given as: $Pu = P_{ct} \wedge P_{mr} \wedge P_{bk}$. The *runs* of the protocol are then defined to be the linear models of Pu. They are all the runs that can be obtained by interleaving the actions of the runs of P_{ct}, P_{mr} and P_{bk}, while synchronizing on common actions. By projecting the runs of the protocol Pu to the alphabets of the participating roles, we get runs of each role P_{ct}, P_{mr} and P_{bk}. As the properties we will consider in this paper regard only the sequence of communicative action exchanged between agents, in the following, we will consider protocol runs as infinite sequences of actions, and disregard worlds.

Note that protocol runs are always infinite, as logic DLTL is characterized by infinite models and we assume that all agents have infinite runs. Therefore, infinite protocols can be easily modelled in our framework. For instance, to model the infinite protocol in which the customer repeatedly issues a request, we can add to the specification above the constraint: $\Box_{ct} \Diamond_{ct} \langle sendRequest \rangle_{ct} \top$, requiring that the message request is issued infinitely many times. When we want to model terminating protocols, as the Purchase protocol above, we assume the domain description of each role of the protocol to be suitably extended with an action $noop_i$ which does nothing and which can be executed forever after termination of the protocol. Hence, in the following, we will assume that, for all i, the i-th projection of a run σ of a protocol P is an infinite run $\sigma|_i$ of P_i.

A protocol specification similar to the one above has been used in [12] to deal with problem of service composition. As a difference, here we use the product version of DLTL so to specify the role of each role P_i independently, while in [12] a global specification of the protocol was given, including all roles.

A protocol defined as above, is not guaranteed to be well behaved, and we must impose some constraints on its structure to guarantee that the different roles participating in the protocol interoperate. To define the interoperability of a set of roles, we will put conditions on the computations of the interacting roles, which can be described by making use of Büchi automata.

2.1   Automata Based Verification

As usual for LTL, a DLTL formula can be mapped into a Büchi automaton so that the accepting runs of the automaton correspond to the models of the formula.

We recall that a *Büchi automaton* has the same structure as a traditional finite state automaton, with the difference that it accepts infinite words. More precisely a Büchi automaton over an alphabet Σ is a tuple $\mathcal{B} = (Q, \rightarrow, Q_{in}, F)$ where:

- Q is a finite nonempty set of states;
- $\rightarrow \subseteq Q \times \Sigma \times Q$ is a transition relation;
- $Q_{in} \subseteq Q$ is the set of initial states;
- $F \subseteq Q$ is a set of accepting states.

Let $\sigma \in \Sigma^\omega$ be an infinite word on Σ. Then a run of $\mathcal{B}$ over σ is a mapping $\rho : prf(\sigma) \rightarrow Q$ such that $\rho(\varepsilon) \in Q_{in}$ and $\rho(\tau) \xrightarrow{a} \rho(\tau a)$ for each $\tau a \in prf(\sigma)$.

The run ρ is *accepting* iff $inf(\rho) \cap F \neq \emptyset$, where $inf(\rho) \subseteq Q$ is given by $q \in inf(\rho)$ iff $\rho(\tau) = q$ for infinitely many $\tau \in prf(\sigma)$. Informally, a run is accepting if it goes infinitely many times trough a final state.

As described in [15], the satisfiability problem for DLTL can be solved in deterministic exponential time, as for LTL. Given a domain description $(\Pi, \mathcal{C})$, a corresponding Büchi automaton can be obtained such that all runs accepted by the automaton represent runs of the protocol, and vice versa. An algorithm for constructing on-the-fly a Büchi automaton from a DLTL formula has been proposed in [10], by generalizing the tableau-based algorithm for LTL. In general, this Büchi automaton is non-deterministic.

Let $P = P_1 \wedge \ldots \wedge P_k$ be a protocol such that each role P_i is specified by a nondeterministic Büchi automaton $\mathcal{M}_i = (Q_i, \rightarrow_i, Q_i^{in}, F_i)$, whose accepting runs provide all the correct executions of the role. We assume that these automata have been "pruned" by eliminating all the states which do not occur on any accepting run. This can be achieved by starting from the accepting states, and by propagating backwards the information on the states for which a path to an accepting state exists.

The interactions of $P_1, \ldots, P_k$ can be described by the runs of a product automaton $\mathcal{M}$ over $\tilde{\Sigma}$. The states of $\mathcal{M}$ will be k-tuples $\langle s_1, \ldots, s_k \rangle$ of states of the automata $\mathcal{M}_1, \ldots, \mathcal{M}_k$. Let $Q^{\mathcal{M}} = Q^1 \times \ldots \times Q^k$ be the set of global states of $\mathcal{M}$. The transitions of $\mathcal{M}$ correspond to the execution of shared actions. The i-local transition relations induce a global transition relation $\rightarrow^{\mathcal{M}} \subseteq Q^{\mathcal{M}} \times \Sigma \times Q^{\mathcal{M}}$ as follows:

$$q \xrightarrow{a}_{\mathcal{M}} q' \text{ iff } q[i] \xrightarrow{a}_i q'[i], \text{ for each } i \in Ag(a)$$

$$q[i] = q'[i], \text{ for each } i \notin Ag(a)$$

where $q[i]$ denotes the ith component of $q = \langle q_1, \ldots, q_k \rangle$, and $Ag(a)$ is the set of agents sharing action a. Moreover, $Q^{in} \subseteq Q_1^{in} \times \ldots \times Q_k^{in}$ is the set of global initial states of $\mathcal{M}$.

A run of $\mathcal{M}$ over $\sigma \in \Sigma^\infty$ is a mapping $\rho : prf(\sigma) \to Q^{\mathcal{M}}$ such that $\rho(\varepsilon) \in Q^{in}$ and $\rho(\tau) \xrightarrow{a}_{\mathcal{M}} \rho(\tau a)$ for each $\tau a \in prf(\sigma)$. The run is *accepting* if, for all $i = 1, \ldots, k$, $\sigma|_i$ is infinite and $\rho(\tau)[i] \in F_i$ for infinitely many $\tau \in prf(\sigma)$. The runs of $\mathcal{M}$ describe the interleaving of the executions of $P_1, \ldots, P_k$, synchronizing on common actions.

Observe that, in an accepting run of $\mathcal{M}$, each $\mathcal{M}_i$ is executing a loop containing at least an accepting state in F_i. We will call such a loop in $\mathcal{M}_i$ an *accepting loop for* P_i. Vice-versa, a loop in $\mathcal{M}_i$ which does not contain any state in F_i will be called a *non accepting loop for* P_i. In the following section we introduce a notion of interoperability of a set of roles $P_1, \ldots, P_k$ by putting conditions on (accepting and non accepting) runs of $\mathcal{M}$ and of the $\mathcal{M}_i$'s.

3   Interoperability

Let us consider an alternative specification P'_{ct} of the customer role, according to which, after sending a request, the customer has to wait for an offer and it does not expect to receive from the merchant the answer "goods is not available". In such a case, the customer role P'_{ct} would not interact properly with the merchant role P_{mr} as defined in the previous section. If the merchant, after receiving a request from the customer, chooses to reply with *sendNotAvail*, the computation gets stuck, as the customer role P'_{ct} cannot receive this message.

This example shows that, as the different roles of the protocol are defined separately, some requirement is needed to guarantee that such roles interact properly, so that the protocol as a whole is well defined. In particular, the interaction of the roles in the protocol should not produce deadlock. Similarly, we want to avoid infinite executions in which some role P_i of the protocol is not executing an accepting run as, either, from some point onwards, P_i does not execute any action, or P_i is executing infinitely many actions, but on a non accepting run for P_i.

In the following, given the roles $P_1, \ldots, P_k$ of a protocol, as introduced in Section 2, we say that roles $P_1, \ldots, P_k$ are *interoperable* when they are free of choosing their actions at each step avoiding deadlock and non accepting executions. Let us point out that here we are considering in two different ways the nondeterministic choices concerning emissions (the customer can accept or refuse an offer) and those concerning receptions (the customer can receive the messages *sendOffer* or *sendNotAvail*). As usual in agent and web service applications, we assume that, in the first case, the choice is internal to the role (internal non determinism), while, in the second case, the choice is external to the agent and depends on the environment, namely on the interleaving of actions of partner agents (external non determinism). Hence, we postulate that a role can choose which message to send among the messages it can send in a state (the customer can decide whether to accept or refuse an offer), but that it cannot chose which message to receive among the messages he can receive in a state (the customer waits for *sendOffer* or *sendNotAvail*, but it cannot choose which one it will receive).

We can now define a notion of interoperability. We will denote by $m(i, j)$ the communicative action m sent from i to j. Let $\pi_i = q_0 \overset{a_1}{\to} q_1 \overset{a_2}{\to} \dots \overset{a_v}{\to} q_v$ be the prefix of a run of $\mathcal{M}_i$. To model the fact that each P_i must be able to choose which action to execute after π_i, we introduce a function $choice(P_i, \pi_i)$, whose value is defined as follows: either $choice(P_i, \pi_i) = m(i, j)$, where $m(i, j)$ is a send action that can be executed after π_i on an accepting run of $\mathcal{M}_i$ (i.e., there is an accepting run of P_i with prefix $q_0 \overset{a_1}{\to} q_1 \overset{a_2}{\to} \dots \overset{a_v}{\to} q_v \overset{m(i,j)}{\to} q_{v+1}$); or the value $choice(P_i, \pi_i) = receiveR$, where $R = \{m_1(j_1, i), \dots, m_n(j_n, i)\}$ contains all the receive actions that can be executed after π_i on an accepting run of $\mathcal{M}_i$. In the last case, P_i expects to receive a message from another agent after π_i but it doesn't know which one it will receive among those messages $\{m_1(j_1, i), \dots, m_n(j_n, i)\}$ it is able to receive after π_i. Observe that the choice of agent P_i in π_i may depend on the state q_v but also on the sequence of actions $a_1, a_2, \dots, a_v$ executed by P_i up to q_v.

While we have assumed that agents can choose among the messages they can send, we have postulated that they cannot decide which message they will receive among those they are able to receive in a given state.

As a matter of notation, in the following, we say that σ is an *execution of* P when there is a (not necessarily accepting) run $q_0 \overset{a_1}{\to} q_1 \overset{a_2}{\to} \dots$ of $\mathcal{M}$ over $\sigma = a_1, a_2, \dots$ (representing a possible computation of $P_1, \dots, P_k$). We say that π is a *finite execution of* P when π is a finite prefix of an execution of P. We say that σ is an *accepting run of* P when there is an accepting run of $\mathcal{M}$ over σ.

We say that σ is an execution of $P_1 \dots P_k$ *respecting the function "choice"* if σ is an execution of $P_1 \dots P_k$ such that, for each prefix $\pi m(i, j)$ of σ, it holds that $choice(P_i, \pi|_i) = m(i, j)$ and $choice(P_j, \pi|_j) = receiveR$, with $m(i, j) \in R$.

We say that *a choice function is fair for* P_i if there is no execution of P_i respecting the choice function, in which P_i executes infinitely many times a non-accepting loop of $\mathcal{M}_i$, although there is a send action $m(i, j)$ that P_i can execute in some state of the loop leading outside the loop. A choice function is *fair* if it is fair for all the P_i's. We say that σ is a *fair execution of* $P_1 \dots P_k$ if σ is an execution of $P_1 \dots P_k$ respecting a fair choice function.

In essence, according to a fair choice function, role P_i cannot choose to execute infinitely many times a non accepting loop, if it can exit the loop by executing a send action: P_i cannot be willing to execute infinitely many times a non-accepting loop. The interoperability of a set of roles is defined as follows:

Definition 1. $P_1 \dots P_k$ *are interoperable if the following conditions hold:*

(i) *For any function* choice *and any finite execution* π *of* $P_1 \dots P_k$*, there exists an action* $m(i, j)$*, such that* $choice(P_i, \pi|_i) = m(i, j)$ *and* $choice(P_j, \pi|_j) = receiveR$ *(with* $m(i, j) \in R$*), and* $m(i, j)$ *is the first action executed after* π *on an execution* σ *of* $P_1 \dots P_k$ *with prefix* π[1].

(ii) *For each fair* choice*, each infinite execution* σ *of* $P_1 \dots P_k$ *respecting* choice*, is an accepting run of* $P_1 \dots P_k$*.*

---

[1] Remember that $\pi|_i$ is the projection of π on the alphabet of P_i.

According to the above definition, any prefix obtained by the execution of $P_1, \ldots, P_k$ can be extended by executing a new action according to the *choice* function. In particular, each agent can choose which action it wants to execute at each stage of the computation and, whatever the choice might be, the computation does not get stuck (condition(i)) and, eventually, each agent can execute his choice in an accepting run (condition (ii)).

Observe that the choice of a role P_i after a sequence π of execution steps only depends on the actions of P_i up to that point (namely, on $\pi|_i$). Hence, the choice of P_i is left unchanged by the execution of communicative actions not involving P_i. As a consequence, the choice of agent P_i remains unchanged until eventually it is fulfilled by the execution of the chosen action.

The conditions above guarantee that the choice of an agent cannot be delayed forever by the choices of other agents. The following proposition follows easily from the definition of interoperability above:

Proposition 1. *(a) For each fair choice functions, for each finite executions π of $P_1 \ldots P_k$, if $choice(P_i, \pi|_i) = m(i,j)$ (or $choice(P_i, \pi|_i) = receiveR$), then there is an accepting run σ of $P_1 \ldots P_k$ respecting the choice function, such that σ has prefix $\pi\pi'm(i,j)$ (respectively, $\pi\pi'm(j,i)$, with $m(j,i) \in R$), where π' does not contain actions of P_i.*
(b) If $P_1 \ldots P_k$ are interoperable, then $P_1 \ldots P_k$ have an accepting run.

Proof. By condition (i), the finite execution π can be extended to an infinite execution σ respecting the fair choice function. By condition (ii) the fair execution σ respecting the fair choice function is an accepting run of $P_1 \ldots P_k$. As P_i must execute infinitely many actions on σ, it will eventually execute an action after π according to his choice in π. Namely, P_i will eventually execute $m(i,j)$ (respectively, $m(j,i)$) as its first action after π. The proof of point (2) is similar.

Let us consider the example in Figure 1. Role P_1 can repeatedly either send message $m(1,2)$ to P_2 or message $m(1,3)$ to P_3. Role P_1 can go on choosing to send $m(1,2)$ to P_2, so that P_3 is not executing any action. This choice is fair. P_2 and P_3 have the only (fair) choice of executing the receive $m(1,2)$ and the receive $m(1,2)$, respectively. P_1, P_2, P_3 are not interoperable. According to the above fair choices, P_1 and P_2 can go on exchanging message $m(1,2)$, and they

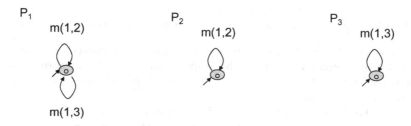

Fig. 1.

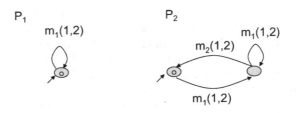

Fig. 2.

produce an execution σ which is an accepting run of P_1 and of P_2, but is not an accepting run of P_3 (P_3 does not execute any action on σ). σ is a fair execution of P_1, P_2, P_3, but it is not a run of P_1, P_2, P_3. Condition (ii) fails.

As a different example, let us consider the example in Figure 2. The runs of P_2 are those sequences obtained by repeating infinitely many times: a (nonempty) finite sequence of actions $m_1(1,2)$ followed by an action $m_2(1,2)$. P_1 and P_2 can go on by exchanging the message $m_1(1,2)$, thus producing a fair execution σ' (in each state, P_2 has the fair choice of executing a receive, but it cannot control which message is received). While for P_1 σ' an accepting run, for P_2 it is not. Although P_2 executes infinitely many actions in σ', it does not execute a run. P_1 and P_2 are not interoperable, as condition (ii) is not satisfied.

4   Conformance

Let $A_1, \ldots, A_k$ be a set of agents. The specification of each agent A_i can be either given through a logical specification as the one introduced in Section 2 for roles, or by introducing the automaton describing the possible behaviors of the agent. We assume that the actions of each agent A_i are deterministic and that the automaton describing its possible behaviors is a deterministic Büchi automaton. The notions of *execution of* $A_1, \ldots, A_k$ and of *accepting run of* $A_1, \ldots, A_k$ are defined as for $P_1, \ldots, P_k$ in Section 3.

Given a protocol $P = P_1 \wedge \ldots \wedge P_k$ with k roles, we define the *conformance of a set of agents* $A_1, \ldots, A_k$ *with* P, as follows:

Definition 2. *Agents* $A_1, \ldots, A_k$ *are* conformant *with P if:*
(a) $A_1, \ldots, A_k$ *are interoperable, and*
(b) all accepting runs of $A_1, \ldots, A_k$ *are accepting runs of the protocol P.*

In this section we want to introduce a notion of *conformance of a single agent* A_i *with the protocol* P, so that the conformance of each A_i, proved independently, guarantees the conformance of the overall set of agents $A_1, \ldots, A_k$ with P, according to Definition 2.

Given the definition of interoperability given in the previous section, this notion of conformance can be based on the policy: *less emissions and more receptions* [4,2,11]. Consider, for instance, a customer agent A_{ct} whose behavior differs from that of the role "customer" of protocol Pu as follows: whenever it receives an offer from the merchant, it always accepts it; after accepting the offer it expects to receive from P_{mr} either *sendGoods* or *cancelDelivery*. Although the behavior of A_{ct} and that of the corresponding role of the protocol are different, we could consider however the agent to be conformant with the protocol, since the customer can choose which messages to send, and thus it is not forced to send all the messages required by the protocol. Also, the agent can receive more messages than those required by the protocol, since these receptions will never be executed.

Let $\mathcal{M}_i$ be the automaton specifying role P_i and $\mathcal{M}_i^A$ the automaton specifying agent A_i. We can formulate the policy "less emissions and more receptions" with the following condition. Let π be a finite execution common to A_i and P_i:

(C1) [Correctness of emissions of A_i]
 If $\pi m(i,j)$ is a finite execution of A_i (where $m(i,j)$ is a send action of A_i), then $\pi m(i,j)$ is a finite execution of P_i.

(C2) [Completeness of receptions of A_i]
 If $\pi m(j,i)$ is a finite execution of P_i (where $m(j,i)$ is a receive action of P_i), then $\pi m(j,i)$ is a finite execution of A_i.

Unfortunately, the policy "less emissions and more receptions" only works for two-party protocols, as shown in the next example.

Example 2. Consider protocol Pu. The customer ct, at some point, may accept the offer of mr and require bk to send the payment to mr. Assume that mr has the requirement that it can receive the payment from bk only after it has received the acceptance of the offer from ct. Namely, the specification of P_{mr} contains the constraint:

(*) $\Box_{mr}(\neg accepted \rightarrow [sendPayment]_{mr}\bot)$.

According to the protocols of ct and bk the message *sendPayment* can be sent from bk to mr either before or after the message *sendAccept* is sent from ct to mr. It is clear that, although ct and bk do not put constraints on the order in which they send the acceptance of the offer and the payment to mr, in the overall protocol Pu they are forced to respect the constraint of the merchant, and only the runs in which *sendAccept* is executed before *sendPayment* are accepted as runs of Pu.

Let us now consider an agent A_{mr}, playing the role of the merchant, whose behavior is the following: either it receives a message *sendAccept* followed by a message *sendPayment*, or receives a message *sendPayment* followed by a message *sendAccept*. Agent A_{mr} allows for more receptions with respect to its role P_{mr} (in fact, A_{mr} receives the additional message *sendPayment* followed by a message *sendAccept*). When A_{mr} interacts with a customer and a bank agents behaving as stipulated by P_{ct} and P_{bk}, it may produce an execution in which *sendPayment* comes before *sendAccept*, which is not a run of protocol Pu (as it does not satisfy the constraint (*) above). Conversely, if we consider a variant Pu' of the Pu protocol in which the roles of the bank P'_{bk} and of the customer P'_{ct} coordinate their executions so that they execute *sendAccept* before *sendPayment*, the interaction of agent A_{mr} with the other roles cannot produce the unwanted execution in which *sendPayment* comes before *sendAccept*.

In this example, A_{mr} can be regarded to be non-conformant with protocol Pu, but conformant with the protocol Pu', although P_{mr} is the same in both protocols. The case in which A_{mr} is non-conformant with Pu, is similar to the example discussed in [17], where the problem of conformance checking is analyzed for models of asynchronous message passing software. The solution adopted in [17] is that of requiring that an agent A_i cannot do more receptions than those established by protocol P_i, so that: A_i can do less emission and exactly the same receptions as stated by P_i. In the example above, this would correspond to take A_{mr} as being non-conformant. We believe that this policy is too restrictive: as we see from Example 2, the conformance of A_{mr} depends on the overall protocol, including other roles. In the following, we propose a definition of the conformance of an agent A_i with respect to the overall protocol P, rather than to its role P_i. We are ready to admit additional receptions in A_i, if we are sure that such receptions cannot give rise to unwanted executions when A_i interacts with agents respecting protocol P.

In the following, besides referring to the executions and runs of a protocol P, we need to refer to the executions and runs of an agent A_i in the context of the protocol P. We will denote by $P[A_i]$ the set of roles/agents $P_1, \ldots, P_{i-1}, A_i, P_{i+1}, \ldots, P_k$. $P[A_i]$ represents the protocol obtained from P by replacing role P_i with agent A_i. Also, we will refer to the executions (accepting runs) of $P_1, \ldots, P_{i-1}, A_i, P_{i+1}, \ldots, P_k$ as executions (accepting runs) of $P[A_i]$.

To guarantee that an agent A_i is conformant with a protocol P, we need to introduce, besides (C1) and (C2) above, further conditions which ensure that A_i interacts properly with the other roles in the protocol P:

(C3) *[Interoperability]* $P[A_i]$ interoperate

(C4) *[Correctness of the receipts of A_i in the context of P]* All accepting runs of $P[A_i]$ are accepting runs of P.

Condition (C3) says that $P[A_i]$ is interoperable, that is A_i interacts with with $P_1, \ldots, P_{i-1}, P_{i+1}, \ldots, P_n$ so that each role can make its choices without the computation getting stuck or ending up in non accepting loops.

Condition (C4) requires that the executions of A_i are *correct* when A_i is interacting with other agents respecting the protocol P. In particular, this ensures that although A_i can execute more receptions than P_i, such additional receptions are not executed when A_i interacts with the other roles in P.

Observe that condition (C4) can be equivalently expressed in the logic by saying that the formula $P[A_i] \to P$ has to be valid.

According to the above definition of conformance, the merchant agent A_{mr} in Example 2 is conformant with Pu', while A_{mr} is non-conformant with protocol Pu. In fact, although $Pu[A_{mr}]$ is interoperable (and, in particular, A_{mr} can interact with other agents executing the protocol without getting stuck), there is a run of $Pu[A_{mr}]$, in which *sendPayment* comes before *sendAccept*, that is not a run of Pu (and therefore (C4) is violated). This last case shows that the interoperability of A_i with the other roles in P is not sufficient to guarantee the correctness of the resulting runs with respect to P.

Are conditions (C1) to (C4) sufficient to guarantee the conformance of an agent with a protocol? What we expect is that, given an interoperable protocol P and a set of agents $A_1, \ldots, A_k$, if each agent A_i is conformant with P (according to (C1)...(C4)) then the agents $A_1, \ldots, A_k$ interoperate and their accepting runs are runs of P.

Consider the roles P_1 and P_2 and the agents A_1 and A_2 in Figure 3. P_1 and P_2 interoperate. A_1 has the same receptions and less emissions than P_1 (the send action $m_3(1,2)$ is not present in A_1). A_1 is conformant with P (according to conditions (C1),...,(C4)). In particular, A_1 and P_2 interoperate: for any fair choice function, P_2 cannot produce the infinite execution $\sigma = m_1(1,2), m_2(2,1)$, $m_1(1,2), m_2(2,1), \ldots$, as P_2 must eventually execute the send action $m_4(2,1)$. Similarly, A_2 is conformant with P and it has the same receptions and less emissions than P_1 (it misses the send action $m_4(2,1)$). However, A_1 and A_2 do not interoperate: A_1 and A_2 have the only choice of executing σ and this is a fair choice for A_1 and A_2. However, σ is not a run of A_1, A_2, and this violates interoperability condition (ii).

Observe that the choice of executing σ is fair for A_1, as A_1 has not the choice of executing any action to exit the non accepting loop in σ. Instead, the execution σ is not fair for P_1, as P_1 can choose to exit the non accepting loop by executing the send action $m_3(1,2)$. We introduce the following condition:

(C5) For each σ execution of both A_i and P_i, if σ is a fair execution of A_i, then σ is a fair execution of P_i.

Condition (C5) is violated by A_1 (and by A_2) in Figure 3. The choice of executing σ is fair for A_1, as there is no send action that agent A_1 can execute to exit from the non accepting loop. Instead, the choice of executing σ is not fair for P_1 which can execute action $m_3(1,2)$ to exit from the non accepting loop. The verification of condition (C5) requires reasoning on non accepting loops in the automata of A_i and P_i. The idea is that, although A_i can contain less

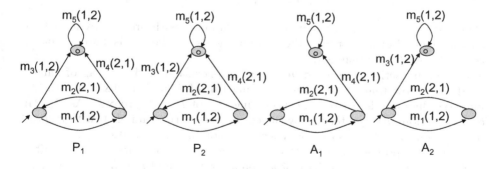

Fig. 3.

emissions than P_i, it must contain al least those emissions allowing to go out from non accepting loop and to go on with an accepting run, if this is possible for P_i.

Finally, we introduce the following conditions to deal with infinite computations, which requires that, when A_i and P_i produce a common action sequence, then either they both accept it or they both do not accept it.

(C6) For all σ that are both executions of A_i and of P_i, σ is an accepting run of A_i if and only if σ is an accepting run of P_i.

The notion of conformance of an agent with a protocol is then defined as follows:

Definition 3. *[Conformance of agent A_i with protocol P] A_i is conformant with a protocol $P = P_1 \wedge \ldots \wedge P_k$ if the conditions (C1),..., (C6) above are satisfied.*

Let $P = P_1 \wedge \ldots \wedge P_k$ be an interoperable protocol. P has at least one accepting run. Let an execution of $A_1, \ldots, A_k$ be a run of the product automaton $\mathcal{A}$ defined as the product automaton $\mathcal{M}$ in Section 2.1. Let an accepting run of $A_1, \ldots, A_k$ be an accepting run of the product automaton $\mathcal{A}$. We can prove that, given k agents $A_1, \ldots, A_k$, if each A_i is conformant with P according to Definition 3, then the accepting runs of $A_1, \ldots, A_n$ are accepting runs of protocol P and $A_1, \ldots, A_n$ interoperate.

Theorem 1. *Let $A_1, \ldots, A_k$ be k agents. If, for each $i = 1, \ldots, k$, agent A_i is conformant with protocol P according to Definition 3, then agents $A_1, \ldots, A_k$ are conformant with P according to Definition 2.*

In order to prove the result above, we prove the following lemmas. Lemma 1 says that any finite execution π of a set of agents $A_1, \ldots, A_k$, with each A_i conformant with P, is a prefix of a run of P.

Lemma 1. *Assume that, for all $i = 1, \ldots, k$, agent A_i is conformant with protocol $P = P_1 \wedge \ldots \wedge P_k$ according to Definition 3. If there is a finite execution π of $A_1, \ldots, A_k$, then there is an accepting run ρ of P with prefix π.*

Proof. We prove that there is an accepting run of P with prefix π by induction on the length l of π. If $\pi = \epsilon$, the theorem holds, as there exists an accepting run of P, by Proposition 1(b).

For the inductive case, let $\pi m(i, j)$ be an execution of $A_1, \ldots, A_k$ of length $l + 1$. We show that there is an accepting run of P with prefix $\pi m(i, j)$. By inductive hypothesis, there is an accepting run of P with prefix π. As A_i can execute action $m(i, j)$ after $\pi|_i$, by the correctness of the send of A_i with respect to P_i, condition (C1), also P_i can execute action $m(i, j)$ just after $\pi|_i$.

From the hypothesis we know that A_j can receive $m(i, j)$ as its first action after π. As P_i can emit $m(i, j)$ as its first action after π, A_j and P_i can synchronize to execute $m(i, j)$ after π in $P[A_j]$. Hence, $\pi m(i, j)$ is a finite execution of $P[A_j]$. Let us consider any fair choice function such that the finite execution $\pi m(i, j)$

respects the choice function. By (C3) $P[A_j]$ interoperate and, by Proposition 1, $\pi m(i,j)$ can be extended to an accepting run ρ of $P[A_j]$ respecting the choice function. By (C4), ρ is also a run of P. Hence, there is an accepting run of P with prefix $\pi m(i,j)$. □

Lemma 1 can be generalized to prove that, if there is a finite execution π of $P[A_1, \ldots, A_h]$ (with $h \leq k$), then there is an accepting run ρ of P with prefix π. We can now prove the following interoperability result.

Lemma 2. *Assume that, for all $i = 1, \ldots, k$, agent A_i is conformant with protocol P according to definition 3. If the roles of P, $P_1 \ldots P_k$, are interoperable, then also $A_1 \ldots A_k$ are interoperable.*

Proof. Assume that $P_1 \ldots P_k$, are interoperable. To show that $A_1 \ldots A_k$ are *interoperable*, we have to prove that $A_1 \ldots A_k$ satisfy conditions (i) and (ii) of Definition 1.

Let us prove (i). Let $choice_A$ be any choice function for $A_1 \ldots A_k$, and let π be any finite execution of $A_1 \ldots A_k$. We want to prove that there exists an action $m(i,j)$, such that $choice_A(A_i, \pi|_i) = m(i,j)$ and $choice_A(A_j, \pi|_j) = receiveR$ (with $m(i,j) \in R$), and $m(i,j)$ can be executed by $A_1 \ldots A_k$ just after π.

By Lemma 1, π is prefix of an accepting run σ of $P_1 \ldots P_k$. Therefore, for all i, P_i has a run $\sigma|_i$ with prefix $\pi|_i$. We want to exploit the fact that $P_1 \ldots P_k$ are interoperable and show that $choice_A$ can be taken as the choice function for the P_i's in π. Let I be the set of all i such that $choice_A(A_i, \pi|_i) = m(i,j)$, that is, all i such that agent A_i wants to execute a send after π.

Let $i \in I$. As $choice_A(A_i, \pi|_i) = m(i,j)$, then A_i can execute $m(i,j)$ after $\pi|_i$ on some of its runs. By the correctness of the send of A_i with respect to P_i, condition (C1), also P_i can choose to execute the send action $m(i,j)$ after $\pi|_i$. Hence, we can take $m(i,j)$ as the choice of P_i after π.

Let $j \notin I$, that is $choice_A(A_j, \pi|_j) = receiveR$. We want to show that also P_j can choose to execute a receive action after π. As π is a finite execution of $A_1 \ldots A_k$ and the prefix of an accepting run of $P_1 \ldots P_k$, then π is a finite execution of $P[A_j]$. By (C3), $P[A_j]$ interoperate. Let us consider any fair choice function ch' for $P_1 \ldots P_k$ such that $ch'(A_j, \pi|_j) = choice_A(A_j, \pi|_j) = receiveR$. By interoperability of $P[A_j]$ there is an accepting run ρ of $P[A_j]$ with prefix $\pi\pi'm(i,j)$, where $m(i,j) \in R$ and π' does not contain actions of agent j. By (C4) ρ is an accepting run of P. As P_i executes the receive $m(i,j)$ as its first action after π in ρ, P_i can choose to execute a receive after π.

Now, let $choice_P$ be a choice function for $P_1 \ldots P_k$, such that $choice_P(P_i, \pi|_i) = choice_A(A_i, \pi|_i)$. By the interoperability of $P_1 \ldots P_k$ there is an action $m(i,j)$ which can be executed by the P_i's after π according to the choice function $choice_P$. Action $m(i,j)$ is the choice of agent A_i after π, so that A_i can send $m(i,j)$ after π. Since P_j receives message $m(i,j)$ after π, by the conformance of A_j to P, condition (C2), also agent A_j can receive message $m(i,j)$ after π. Hence, $m(i,j)$ can be executed by $A_1 \ldots A_k$ after π according to the choice function $choice_A$. This concludes the proof of point (i).

Let us prove (ii). We have to prove that, for each fair *choice*, each infinite execution σ of $A_1, \ldots, A_k$ respecting *choice*, is an accepting run of $A_1, \ldots, A_k$.

Let σ be an infinite execution of $A_1, \ldots, A_k$ respecting *choice*. It is easy to see that σ is also an execution of P. In fact, by Lemma 1 we know that each prefix π of an execution of $A_1, \ldots, A_k$ is also a prefix of a run of P. Hence, σ is also an execution of $P_1, \ldots, P_k$. Given condition (C5), the *choice* function, which is fair for $A_1, \ldots, A_k$, must also be a fair choice function for $P_1, \ldots, P_k$. As P is interoperable and the *choice* function is fair for $P_1, \ldots, P_k$, by the interoperability condition (ii), σ is an accepting run of P. Observe that, for all i, $\sigma|_i$ is both an execution of A_i and of P_i. By condition (C6), as $\sigma|_i$ is an accepting run of P_i, $\sigma|_i$ must also be an accepting run of A_i. It follows that σ is an accepting run of $A_1, \ldots, A_k$.   □

Lemma 3. *Assume that, for all $i = 1 \ldots k$, agent A_i is conformant with protocol P according to definition 3. If σ is an accepting run of $A_1 \ldots A_k$, then σ is an accepting run of P.*

Proof. By induction on h, we prove that: If σ is an accepting run of $P[A_1, \ldots, A_h, A_{h+1}]$, then σ is an accepting run of $P[A_1, \ldots, A_h]$. For $h = 0$, by (C4), if σ is an accepting run of $P[A_1]$, then σ is an accepting run of P.

For $h > 0$, assume that σ is an accepting run of $P[A_1, \ldots, A_h, A_{h+1}]$ and that σ is not an accepting run of $P[A_1, \ldots, A_h]$. It must be that $\sigma|_i$ is an accepting run of A_{h+1}, while $\sigma|_i$ is not an accepting run of P_{h+1}.

Observe that, by (C1), all the send actions of A_{h+1} in σ are correct and can be executed by P_{h+1}, but A_{h+1} could receive a message $m(j, h+1)$ in σ which cannot be received by P_{h+1}. If this is not the case, σ is an execution of $P[A_1, \ldots, A_h]$ and, by (C6), σ is an accepting run of $P[A_1, \ldots, A_h]$.

If there is a receive $m(j, h+1)$ in σ which is not executable by P_{h+1}, then there must be a prefix $\pi m(j, h+1)$ of an execution of $P[A_1, \ldots, A_h, A_{h+1}]$ which is not a prefix of an accepting run of P (as it is not a prefix of an accepting run of P_i). This contradicts Lemma 1.   □

Observe that, although the interoperability of A_i with P (condition (C3)), guarantees that A_i is able to receive the messages sent to it from the other roles in P, it does not enforce condition (C2). If (C2) were omitted in Definition 3, Theorem 1 would not be provable. In fact, in such a case, it might occur that, although agent A_1 is interoperable with other roles in P and the same holds for agent A_2, the two agents A_1 and A_2 do not interoperate with each other and with the other roles in P.

Let, for instance, P be an interoperable protocol with 4 roles, where each role has the following runs:

$$P_1 : \{m(1,3), m(1,4)\}$$
$$P_2 : \{m(2,4), m(2,3)\}$$
$$P_3 : \{m(1,3), m(2,3) + m(2,3), m(1,3)\}$$
$$P_4 : \{m(1,4), m(2,4) + m(2,4), m(1,4)\}$$

Let us consider now the agents A_1, A_2, A_3, A_4, such that $A_1 = P_1$, $A_2 = P_2$, while A_3 and A_4 have the following runs:

$$A_3 : \{m(2,3), m(1,3)\} \qquad A_4 : \{m(1,4), m(2,4)\}$$

It can be seen that A_3 satisfies conditions (C1) and (C3), but not (C2), and the same holds for A_4. The agents A_1, A_2, A_3, A_4 do not interoperate, since A_1 and A_2 must emit respectively $m(1,3)$ and $m(2,4)$ as their first message, but none of them can be received by A_3 or A_4.

Observe that some conditions in the definitions of interoperability and conformance can be easily verified, as for instance, the interoperability condition (i), which requires to check some conditions on all the states of the product automaton: from each state it must always be possible to execute some action according to the choices of the agents. This verification requires polynomial time in the size of the product automaton $\mathcal{M}$. However, the verification of other conditions, like the interoperability condition (ii) and of conditions (C4) and (C5) is rather complex. For instance, the verification of condition (ii) requires to execute a check on all loops in the automaton $\mathcal{M}$, that are non accepting loops for some role P_i. Also, the verification of condition (C4), that all the runs of $P[A_i]$ are runs of P, requires to check language inclusion between two non deterministic ω automata. This problem is, in the general case, PSPACE-hard [7].

We may wonder whether the addition of simplifying assumption may lead to a simplified definition of the notions of conformance and interoperability. Several simplifications naturally arise: the assumption that the agents have the same receptions as the corresponding roles; the assumption that the role and agent automata are deterministic; the assumption that all protocol are finite and that finite automata can be used rather then Büchi automata to model protocols.

Concerning the first simplification, if agent A_i has that same receptions as P_i, conditions (C1) and (C3) can be replaced by the single condition (C1') All the runs of A_i are runs of P_i stating that both emissions and receptions of A_i are correct with respect to P_i. Verifying this condition, requires to verify the validity of the entailment $A_i \rightarrow P_i$, which for DLTL (as for LTL) is a problem in PSPACE-hard. With this simplification however condition (C5) is still needed to guarantee interoperability, as shown, for instance by the example in Figure 3.

If we restrict our consideration to protocols described by deterministic Büchi automata, rather than to non deterministic ones, the verification of some condition becomes simpler (as, for instance, the verification of (C4)), although the conditions for interoperability and conformance remain unaltered.

In the case we are considering finite state automaton, some conditions, like (C6), are not needed. Also, the interoperability conditions (i) and (ii) gets simplified. However, to guarantee interoperability of the agents which are individually conformant with the protocol, some condition playing the role of (C5) is still needed, as shown by the example in Figure 3, which can be easily adjusted for finite state automata.

5   Related Work

The paper deals with the problem of agents conformance with multiparty protocols. The notion conformance we have introduced guarantees the interoperability of a set of agents which are conformant with the protocol. We have assumed that the specification of the protocol is given in a temporal action logic and we have shown that the verification of conformance can be done independently for each agents, by making use of automata-based techniques. The proposed approach deals with both terminating and non-terminating protocols.

The paper generalizes the proposal in [11], where a notion of conformance for two-party protocols has been defined based on the policy "less emissions and more receptions". Such policy, however, is not sufficient to guarantee stuck-freeness in the multiparty case. Indeed, the notion of conformance we propose here requires in addition: 1) the correctness of the receipts an agent can do, when it is interacting with other roles of the protocol, and 2) the interoperability of the agent with the other roles of the protocol. The notion of conformance proposed here is also stronger that the notion of compliance in [13]. There, an agent A_i is said to be compliant with a protocol P if, in all interactions of A_i with other roles in the protocol, A_i satisfies its commitments and permissions. This condition essentially corresponds to the correctness condition (C2) in Definition 3. The notion of compliance in [13] does not guarantee stuck-freeness.

Several other proposals have been put forward in the literature for dealing with agent and agent conformance and interoperability.

In [4], several notions of *compatibility* and *substitutability* among agents have been analyzed, in which agents are modelled by Labelled Transition Systems, communication is synchronous, and models are deterministic. Substitutability is related to the notion of conformance. [4] introduces two distinct notions of substitutability (related with conformance): a first one, based on the policy "less emissions and more receptions", which does not preserves deadlock-freeness, and a second more restrictive one requiring "the same emissions and receptions". Agent executions are always terminating.

In [2] an automata based approach is used for conformance verification, by taking into account the asymmetry between messages that are sent and messages that are received. Agents and protocols are represented as deterministic finite automata, and protocols have only two roles. The approach has been extended to the multiparty case in [3], which also accounts for the case of nondeterministic agents and roles producing the same interactions but having different branching structures. Such a case cannot be handled in the framework in [2] as well as in our framework, due to the fact that our approach is based on a trace semantics. A similar approach is also used in [1], where an abductive framework is used to verify the conformance of agents to a choreography with any number of roles. As a difference with the above proposals, our proposal deals with protocols with infinite runs and guarantees stuck-freeness also in the multiparty case.

In [17] a notion of conformance is defined to check if an implementation model I conforms with a signature S, in the case both I and S are CCS processes, and communication is asynchronous. The policy "less emissions and the same receptions" is introduced to guarantee stuck-freeness. Our approach provides a solution to guarantees stuck-freeness without requiring an agent to have the same receptions as its role in the protocol.

The notions of conformance, coverage and interoperability are defined in [5]. A distinctive feature of that formalization is that the three notions are orthogonal to each other. Conformance and coverage are based on the semantics of runs and relies on the notion of run subsumption, concerning the single agent and its role in the protocol. Interoperability among agents is based upon the idea of blocking and depends on the computation that the agents can jointly generate. The paper only considers two-party protocols and agents with finite runs.

In [6] a notion of constitutive interoperability is proposed, which "abstracts from the process-algebraic notion of interoperability and makes commitment alignment the sole criteron" so to capture the business meaning of the business processes. As said in [6], this notion is complementary to a notion of regulative interoperability, which takes into consideration message order, occurrence and data flow. The work we have presented here falls within the context of regulative interoperability.

In [16] Web Services are modelled as MAS and model checking is used for verifying temporal epistemic properties of OWL specifications. An OWL specification of a service is mapped to ISPL (the language of the model checker MCMAS), and a coloring of states as compliant (green) or non compliant (red) is assumed for the verification. The paper does not provide formalization of contracts (or protocols) and it does not provide techniques for automatically computing, for given a service, the states which are compliant or non compliant with a contract.

[8] focuses on the realizability problem of a framework for modeling and specifying the behaviors of *reactive electronic services*. In that framework, services communicate by asynchronous message passing, and are modeled by means of Büchi automata. The authors show that not every conversation protocol is realizable in the framework, and give some realizability conditions and show that each conversation protocol satisfying those conditions is realizable.

Acknowledgement

The work was partially supported by Regione Piemonte, Project ICT4LAW.

References

1. Alberti, M., Chesani, F., Gavanelli, M., Lamma, E., Mello, P., Montali, M.: An abductive framework for a-priori verification of web agents. In: Principles and Practice of Declarative Programming (PPDP 2006). ACM Press, New York (2006)
2. Baldoni, M., Baroglio, C., Martelli, A., Patti, V.: Verification of protocol conformance and agent interoperability. In: Toni, F., Torroni, P. (eds.) CLIMA 2005. LNCS, vol. 3900, pp. 265–283. Springer, Heidelberg (2006)
3. Baldoni, M., Baroglio, C., Martelli, A., Patti, V.: A Priori Conformance Verification for Guaranteeing Interoperability in Open Environments. In: Dan, A., Lamersdorf, W. (eds.) ICSOC 2006. LNCS, vol. 4294, pp. 339–351. Springer, Heidelberg (2006)

4. Bordeaux, L., Salaün, G., Berardi, D., Mecella, M.: When are two web-agents compatible, VLDB-TES (2004)
5. Chopra, A.K., Singh, M.P.: Producing Compliant Interactions: Conformance, Coverage, and Interoperability. In: Baldoni, M., Endriss, U. (eds.) DALT 2006. LNCS, vol. 4327, pp. 1–15. Springer, Heidelberg (2006)
6. Chopra, A.K., Singh, M.P.: Constitutive Interoperability. In: Proc. of the 7th Conf. on Autonomous Agents and Multiagent Systems (AAMAS 2008), pp. 797–804 (2008)
7. Clarke, E.M., Grumberg, O., Peled, D.A.: Model Checking. MIT Press, Cambridge (2000)
8. Fu, X., Bultan, T., Su, J.: Conversation protocols: a formalism for specification and verification of reactive electronic services. Theor. Comput. Sci. 328(1-2), 19–37 (2004)
9. Giordano, L., Martelli, A., Schwind, C.: Verifying Communicating Agents by Model Checking in a Temporal Action Logic. In: Alferes, J.J., Leite, J. (eds.) JELIA 2004. LNCS (LNAI), vol. 3229, pp. 57–69. Springer, Heidelberg (2004)
10. Giordano, L., Martelli, A.: Tableau-based Automata Construction for Dynamic Linear Time Temporal Logic. Annals of Mathematics and Artificial Intelligence 46(3), 289–315 (2006)
11. Giordano, L., Martelli, A.: Verifying Agent Conformance with Protocols Specified in a Temporal Action Logic. In: Basili, R., Pazienza, M.T. (eds.) AI*IA 2007. LNCS (LNAI), vol. 4733, pp. 145–156. Springer, Heidelberg (2007)
12. Giordano, L., Martelli, A.: Web Service Composition in a Temporal Action Logic. In: 4th Int. Workshop on AI for Service Composition, AICS 2006, Riva del Garda, August 28 (2006)
13. Giordano, L., Martelli, A., Schwind, C.: Specifying and Verifying Interaction Protocols in a Temporal Action Logic. Journal of Applied Logic (Special issue on Logic Based Agent Verification) 5(2007), 214–234 (2007)
14. Henriksen, J.G., Thiagarajan, P.S.: A product Version of Dynamic Linear Time Temporal Logic. In: Mazurkiewicz, A., Winkowski, J. (eds.) CONCUR 1997. LNCS, vol. 1243, pp. 45–58. Springer, Heidelberg (1997)
15. Henriksen, J.G., Thiagarajan, P.S.: Dynamic Linear Time Temporal Logic. Annals of Pure and Applied logic 96(1-3), 187–207 (1999)
16. Lomuscio, A., Qu, H., Solanki, M.: Towards verifying compliance in agent-based web service compositions. In: AAMAS 2008, pp. 265–272 (2008)
17. Rajamani, S.K., Rehof, J.: Conformance checking for models of asynchronous message passing software. In: Brinksma, E., Larsen, K.G. (eds.) CAV 2002. LNCS, vol. 2404, pp. 166–179. Springer, Heidelberg (2002)
18. Reiter, R.: Knowledge in Action. MIT Press, Cambridge (2001)
19. Reiter, R.: The frame problem in the situation calculus: a simple solution (sometimes) and a completeness result for goal regression. In: Lifschitz, V. (ed.) Artificial Intelligence and Mathematical Theory of Computation: Papers in Honor of John McCarthy, pp. 359–380. Academic Press, London (1991)
20. Singh, M.P.: A social semantics for Agent Communication Languages. In: Dignum, F.P.M., Greaves, M. (eds.) Issues in Agent Communication. LNCS, vol. 1916, pp. 31–45. Springer, Heidelberg (2000)
21. van der Aalst, W.M.P., Pesic, M.: DecSerFlow: Towards a Truly Declarative Service Flow Language. In: Bravetti, M., Núñez, M., Zavattaro, G. (eds.) WS-FM 2006. LNCS, vol. 4184, pp. 1–23. Springer, Heidelberg (2006)
22. Yolum, P., Singh, M.P.: Flexible Protocol Specification and Execution: Applying Event Calculus Planning using Commitments. In: AAMAS 2002, Bologna, Italy, pp. 527–534 (2002)

Run-Time Semantics of a Language for Programming Social Processes*

Juan M. Serrano and Sergio Saugar

University Rey Juan Carlos
C/Tulipan S/N
Madrid, Spain
juanmanuel.serrano@urjc.es, sergio.saugar@urjc.es

Abstract. There is a broad range of application domains which can be described under the heading of social process domains: business processes, social networks, game servers, etc. This paper puts forward a programming language approach to the formal specification of social processes, building upon the C+ action description language. Particularly, the paper focuses on the run-time semantics of the language, which is delivered as a core layer of application-independent sorts which make up the abstract machine of the language. The advantages of the presented approach with respect to other state-of-the-art proposals lie in its strong support for modularity and reusability, and hence for the development of large-scale, elaboration tolerant, specifications of social processes.

1 Introduction

Social processes refer to any kind of joint activity performed by humans within a given social context. From this perspective, social processes can be regarded as one of the key elements of a wide class of application domains. For instance, they are at the core of *business process* applications, in the form of operational (e.g. inventory, sales) or managerial processes; they pervade *online communities* and *social software* applications in the web 2.0 realm, where wikis, discussion forums, chats, etc., serve as vehicles for everyday communication; *online game servers* represent another significant domain, for games themselves can be understood as social interactions; last, social processes are ubiquitous in e-administration, e-government, e-democracy, etc., application domains. Social process applications thus refer to any software application which is primarily designed to support human interaction, irrespective of the kind of setting within which it happens (business, leisure, political, administrative, etc.).

The approach to the development of social process applications advocated in this paper rest upon three major premises. Firstly, a social process application shall be delivered as a *computational society*, viz. a distributed system where a social middleware (e.g. AMELI [10], S-Moise+ [12]) provide heterogeneous and autonomous software components with high-level interaction mechanisms

---

* Research sponsored by the Spanish MICINN, project TIN2006-15455-C03-03.

M. Fisher, F. Sadri, and M. Thielscher (Eds.): CLIMA IX, LNAI 5405, pp. 37–56, 2009.

(e.g. communicative actions, dialogue games, scenes, organisations, institutions). These mechanisms commonly rely on normative protocols, based on concepts such as empowerments, permissions, contracts, commitments, and so forth. Thus, computational societies contribute with high-level software abstractions which fit the required level of expressiveness and domain adequacy. Secondly, social processes shall be implemented as first-class *software connectors* [14], i.e. as a particular kind of interaction mechanism. Thus, the implementation of social processes is completely decoupled from the development of the software components that will engage in those processes at run-time. This separation of concerns between interaction and computation is generally rewarded with an improvement in reusability, maintainability and evolvability. Last, we advocate a *programming language* stance on the development of social process applications. Essentially, this perspective resolves itself into two major aspects: firstly, the social middleware shall be regarded as a *programmable machine*; secondly, the middleware behaviour shall be programmed through a collection of social connector *types*. These types encode the rules and structure of the social process domain in the form of a computational society specification to be interpreted by the social middleware.

This paper takes some preliminary steps towards the design of a language for programming social processes by addressing its run-time semantics, i.e. the specification of the social middleware infrastructure to be programmed. At the core of this specification is an appropriate notion of social interaction which is able to capture the major computational aspects of social processes [17]. Also, the proposed specification goes along in major aspects with other proposals to the specification and programming of computational societies or normative multi-agent systems [6,5,2]. We will argue, however, in favour of our approach on grounds of its support for key principles in the design of large-scale specifications, such as modularity, reusability and separation of concerns.

The specification of the run-time semantics will be formalised using the C+ action language [11], a declarative formalism originated in the artificial intelligence field for specifying and reasoning about dynamic domains. The semantics of a C+ action description is given in terms of transition systems, hence the operational flavour of the proposed specification. The rest of the paper is structured as follows. Firstly, the major features of the C+ language and its accompanying CCALC interpreter will be reviewed. Then, the basic postulates which guided the application of the C+ action language to our problem will be presented. The next three sections present the action description of the social middleware dynamics structured according to the major sorts of institutional entities – social interactions, agents and social actions. The paper finishes with a small example in the conference management domain and a discussion on the advantages and limitations of the current specification.

2   Review of the Action Language C+

The action language C+ [11] is a declarative formalism for describing the behaviour of dynamic domains. C+ action descriptions consist of a collection of *causal laws* defined over a (multi-valued) propositional signature of *fluent* (or state parameters) and *action* constants (or events). As any action language [3], the semantics of a C+ action description D is formalised as a *labeled transition system* T_D. The nodes of T_D represent possible states of the domain; its arcs, the one-step evolution of the domain due to the concurrent execution of several actions (possibly, none). States $s, s', \ldots$ of T_D are complete and consistent interpretations of the signature's fluent constants. Similarly, each transition (s, l, s') of T_D is labeled with a complete and consistent interpretation l of the action constants. The actions executed in a transition are, precisely, those that hold in l. There are three types of causal laws: *static laws, action dynamic laws* and *fluent dynamic laws*.

Static laws are causal laws of the form **caused** F **if** G, where both F and G are fluent formulas. Intuitively, these laws establish that if G holds in a state s then there is a cause for the head of the law, F, to hold in s as well. Thus, they establish a causal dependency between the fluent constants of G and F *in the same state*.

The labels of transitions (i.e. the actions to be executed) are determined through so-called *action dynamic laws*, viz. causal laws of the form **caused** F **if** G, where F is an action formula and G a general formula (i.e. possibly containing both action and fluent constants). If G is a fluent formula which holds in a state s, these laws establish that there is a cause for F to hold (i.e. be executed) in any leading transition (s, l, s'); alternatively, if G is an action formula which holds in a transition t then there is a cause to execute F in t as well.

The values of fluent constants at the destination state of a transition are determined through so-called *fluent dynamic laws*, viz. causal laws of the form **caused** F if G **after** H, where F and G are fluent formulas and H a general formula. If H is a fluent formula, these laws establish that there is a cause for F to hold in a state s' of any transition (s, l, s') if G also holds is s' and H holds in the source state s; if H is an action formula, they establish that there is a cause for F to hold in s' provided that G also holds in s' and H holds in l (i.e. is executed). Thus, these laws allow to represent both the effects of actions[1] and the law of inertia, as indicated bellow.

In order to simplify action descriptions, the C+ language defines a collection of abbreviations for common patterns of causal laws [11, Appendix B]. The following abbreviations are specially relevant for this paper:

- **default** F, where F is a formula. This expression establishes that there is a cause for F to hold in a state or transition, if it is indeed consistent to assume that it holds.
- **inertial** c, where c is a fluent constant. It means that for any transition (s, l, s'), the value of constant c in s' is by default the same as in state s.
- **exogeneous** c, where c is an action constant. Intuitively, exogenous actions represent a particular class of actions whose cause for execution is to be found outside the domain being modeled.

---

[1] Indirect effects of actions, i.e. ramifications, may then be represented with static laws.

- **constraint** F, where F is a fluent formula. These laws filter any interpretation s from the states of the transition system in which F holds.
- F **causes** G **if** H, where F is an action formula. It establishes that actions F of a transition (s, l, s') have an effect G in s' provided that H holds in s. The part "**if** H" can be dropped if H is $\top$ (in that case the effect is unconditional).
- **nonexecutable** F **if** G, where F is an action formula and G a formula. It filters any transition (s, l, s') where F is executed at the same time that G holds in s and/or l. Intuitively, it allows to represent preconditions of actions in reference to certain properties of states and/or concurrently executing actions.

The *Causal Calculator* (CCALC)[2] is an interpreter of C+ which allows to query the transition systems of *definite* action descriptions [11, sec. 5] for specific paths (or histories) of a given maximum length. For instance, if queried for any possible path of length 0 (resp. length 1), the interpreter returns all states of the transition system (resp. all transitions). In general, by constraining the values of fluents and action constants at different time instants of paths, it is possible to define different planning, prediction and postdiction queries. CCALC is a tool implemented in Prolog that finds the answer to a query by translating it to a propositional theory, which is then passed to a SAT solver.

The input of CCALC is a Prolog file which encodes the signature and causal laws of an action description. The major features of the input language of CCALC are the following [11, sec. 6][1, sec. 3]:

- *Sorts, objects* and *variables.* To make easier the specification of action descriptions, CCALC allows to specify a collection of *sorts* of entities, together with their domains (i.e. *objects* of the corresponding sorts) and variables. Sorts can be used in constant declarations as placeholders for their corresponding domains. Similarly, variables of the declared sorts can be used in the declaration of causal laws in place of their corresponding objects. Moreover, sorts can be arranged in a hierarchy of *subsorts* through declarations of the form `:- sorts a >> b`. This expression declares two sorts `a` and `b` and establishes that every object of the subsort `b` is also an object of the supersort `a`. The actual declarations of constants and causal laws are obtained through a grounding mechanism that takes the above declarations into account.
- *Constant declarations.* Simple and statically determined fluent constants are declared through the keywords `simpleFluent` and `sdFluent` respectively. The special keyword `inertialFluent` allows to declare a simple fluent which is also inertial, without requiring the programmer to explicitly include the corresponding causal law. For instance, the expression `:- constants f ::` `inertialFluent(a)` declares a simple fluent f whose values range in the domain of sort a (in case that a is the boolean sort it may be omitted from the declaration) and implicitly declares the causal law **inertial** f. Similarly,

---

[2] `http://www.cs.utexas.edu/users/tag/cc/`

exogeneous actions can be declared with the `exogeneousAction` keyword. Another facility is the possibility of declaring *partial valued* fluent constants. For instance, the expression `:- constants f :: sdFluent(a+none)` declares a statically determined fluent f whose values range in the domain of sort a, plus the special object `none`.

- *Causal law declarations.* CCALC implements all the abbreviations for causal laws defined in [11].
- `Include` *directives.* CCALC provides an `include` directive which imports into the current file, the sorts, objects, constants, variables and causal laws declared elsewhere. This allows to distribute the implementation of a complex action description D through a collection of specifications D_a, D_b, etc., stored in different files.

3   Social Middleware as a C+ Action Domain

In applying the C+ action language to our problem we must first of all identify the dynamic domain to be modeled, viz. the *social middleware infrastructure* (i.e. the abstract machine which manages the computational society). Thus, the software components that communicate through this kind of middleware are external, non-modeled entities. The following guidelines and considerations have been postulated for the specification of the action description:

- The major types of entities of a computational society will be represented as different sorts. This paper focuses on *social interactions* ($\mathcal{I}$), *agents* ($\mathcal{A}$), *social actions* (Act) and, to a lesser extent, *environmental resources* ($\mathcal{R}$). Other entities such as *invocations* of services provided by computational resources, *obligations* of agents, etc., are outside the scope of this paper.
- The sets of fluents, objects, actions, variables and causal laws of the whole action description, named D_{Speech}, will be partitioned according to the previous sorts into the action descriptions $D_{\mathcal{I}}$, $D_{\mathcal{A}}$, $D_{\mathcal{R}}$ and D_{Act}. These specifications describe the structure and dynamics shared by any kind of social interaction, agent, etc.[3]
- The actions performed by external software components over the middleware, viz. *entering* the society as an agent, *attempting* their agents to do something, *exiting* the society, etc., will be represented as exogenous C+ actions. This paper exclusively focuses on the *attempt* external action.
- Non-exogenous C+ actions shall represent the internal actions executed by the middleware to update the state of the computational society (e.g. to *initiate* a social interaction). These internal actions may be executed in response to an external action (e.g. an attempt), or automatically when the computational society reaches some state (e.g. the middleware may be programmed to automatically initiate some interaction when certain conditions hold).

---

[3] Thus, the generic sorts $\mathcal{I}$, $\mathcal{A}$, etc., play a similar role to the Java `Object` class.

42   J.M. Serrano and S. Saugar

The action description D_{Speech} specifies the generic, application-independent behaviour of the social middleware infrastructure. To program this behaviour in accordance with the requirements of a particular social process application, an action description D_{appl}, composed of a collection of application-dependent sort specifications, must be provided[4]. These sorts are declared through the CCALC *subsort* mechanism by extending the generic sorts $\mathcal{I}$, $\mathcal{A}$, etc. Thus, they extend the collection of *standard* constants and causal laws, inherited from the generic specifications, with new application-dependent action and fluent constants, variables and causal laws, etc. The precise way in which this is accomplished, however, is addressed by the type system of the language. This component is outside the scope of this paper, which just focuses on the common structure shared by any kind of computational society and the behaviour of the social middleware infrastructure in charge of their management, i.e. on the run-time semantics of the programming language, as specified by the action description D_{Speech}. This language is named SPEECH in recognition of the role played by communicative actions (a particular kind of social action, as discussed in section 6) as one of the major interaction mechanisms offered by the social middleware.

To illustrate this specification, the following sections will refer to the run-time snapshot of a computational society for conference management shown in figure 1. In this figure, social interactions instances are represented by round corner rectangles, agents by stick figures and resources by triangles; last, speech bubbles represent the performance of social actions.

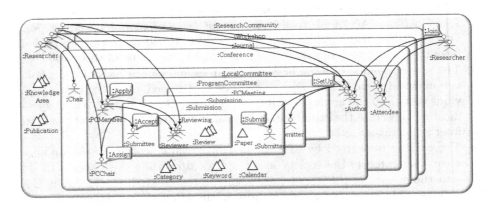

Fig. 1. Run-time snapshot of a multiagent society for conference management

---

[4] It should be noted that, in general, the descriptions D_{Speech} and D_{appl} abstract away from any particular domain, i.e. the extents of sorts are not declared. Thus, in order to define a working transition system (i.e. one which can be queried), both specifications must be complemented with a description D_{domain} of the domain in question.

4   Social Interactions

In accordance with the middleware perspective endorsed by this paper, social interactions will be regarded as the major building blocks of computational societies. In particular, a computational society shall be formalised as an interaction space hierarchically structured in terms of a tree of nested social interactions. The root of this tree, or top-level interaction, represents the computational society itself. Thus, the whole activity within a social interaction can be decomposed into different *sub-interactions*, and every social interaction, but the top-level one, takes place within the *context* of some other interaction. Any social interaction represents the context within which agents carry out their activities. The population of agents deployed within the context of a particular interaction represents its *member* agents. Also, social interactions provide the context for its set of *environmental* resources, which agents create and manipulate. Both, agents and environmental resources, represent the two kind of social roles played by software components within the computational society [17].

For instance, as shown in figure 1, the *research community* is the top-level interaction of the computational society designed to support conference management. This interaction provides the context in which conferences take place as well as other kinds of processes such as workshops, journals, etc.[5] The whole activity within conferences is decomposed into different sub-interactions such as the *local* and *program committees* (PCs), which are complex processes as well. Thus, the program committee is further decomposed into different *submissions* processes (one for each paper to be submitted), the *PC meeting* where the final decision will be taken on the acceptance and rejection of papers, etc. Following the branch of submissions, the *reviewing* process encapsulates the activity related with the evaluation of its submission context; this interaction represents a leaf of the sub-interaction tree, i.e. an atomic social interaction. As far as agents and resources are concerned, the members and environmental resources of every interaction in figure 1 are those placed within their limits. Agents will be described in the next section. Concerning resources, figure 1 illustrates different kinds of informational resources related with the management of conferences: at the level of the research community, we find the *knowledge areas* which represent the research fields of the community, and the *publications* (e.g. papers of conferences, articles of journals, technical reports, etc.) in which the knowledge of the community is encoded; the program committee is in charged of defining the *categories* and *keywords* for paper submission, as well as the *calendar* which sets the important dates (deadlines for submission, notification, reviewing, etc.). The two other kinds of resources illustrated in figure 1 are *papers*, within the context of submissions, and *reviews* created within the context of the reviewing process.

---

[5] In a more realistic scenario, the context of conferences wouldn't be the research community itself, but the *conference series*. Besides the different *conference editions*, this interaction would provide the context for the *steering committee* as well.

The features of social interactions discussed so far (sub-interactions, member agents and environmental resources) can be characterised as *structural*. On the other hand, the *run-time state* of interactions, which could be *open* or *closed*, is a dynamic feature. An open interaction represents an ongoing process, whereas a closed interaction represents a finished one. Interactions may be *initiated* and *finished* in two major ways. Firstly, through the performance of the standard social actions *set up* and *close*, as described in section 6; and, secondly, when certain conditions hold or events occur. For instance, a conference edition may be finished automatically when the presentation sessions are over or when the conference chair *closes* it prematurely (e.g. because of a small number of submissions).

The above features postulated for any kind of social interaction are formally specified by the C+ generic interaction sort $\mathcal{I}$, whose related specification $D_{\mathcal{I}}$ is shown in figure 2. This signature includes the auxiliary sort $\mathcal{S}_{\mathcal{I}}$ and the objects *open* and *closed*, which represent the possible run-time states of interactions. The specification also includes the predefined *top* interaction, which represent the top-level social interaction – the only mandatory interaction of any computational society. The following paragraphs explain the most salient characteristics of the formalisation.

States. The inertial fluent *state* represents the run-time execution state of social interactions. Besides *open* and *closed*, the run-time state of some interaction may be *none*. This value allows to represent "non-existent" interactions, i.e. interactions which have not been institutionally *initiated* yet, and constitute the pool of available interaction objects[6]. The major structural features of social interactions are represented by the inertial fluents *member* (which refer to the agent sort $\mathcal{A}$), *environmental* resources (which are represented by sort $\mathcal{R}$) and *sub*-interactions. These fluents are complemented by the *statically determined* fluents *context*, a partial non-boolean fluent representing the possible context of the interaction; and sub_+, a boolean fluent representing the transitive closure of the sub-interaction relationship. These fluents, defined by static laws 1–4, are introduced to facilitate the declaration of other causal laws.

According to causal law 5, non-existent interactions can neither hold subinteractions, member agents or environmental resources, nor be part of some interaction context. Causal laws 6 and 7 define the interaction space of a computational society as a sub-interaction tree whose root is the predefined *top* interaction object. The run-time state of the interactions of this tree must satisfy laws 8 and 9, which establish that the top-level interaction of the society is always open; and that some interaction can not be open if its context is closed. Moreover, the static law 9 also establishes a causal relationships between the states of a sub-interaction and its context, as explained bellow.

---

[6] This pool is needed in order to comply with the *domain closure assumption* [15] endorsed by the C+ language. This assumption establishes that all objects in the domain of discourse must be explicitly declared in advance by constants of the signature.

:- **sorts**
  $\mathcal{I}$; $\mathcal{S_I}$.
:- **objects**
  $top :: \mathcal{I}$;
  $open,\ closed :: \mathcal{S_I}$.
:- **constants**
  $state(\mathcal{I}) ::$ **inertialFluent**$(\mathcal{S_I}+none)$;
  $member(\mathcal{I},\mathcal{A})$, $env(\mathcal{I},\mathcal{R})$, $sub(\mathcal{I},\mathcal{I}) ::$ **inertialFluent**;
  $context(\mathcal{I}) ::$ **sdFluent**$(\mathcal{I}+none)$;
  $sub_+(\mathcal{I},\mathcal{I}) ::$ **sdFluent**;
  $initiate(\mathcal{I},\mathcal{I})$, $finish(\mathcal{I}) ::$ **action**.
:- **variables**
  $i,\ i_c,\ i_{cc} :: \mathcal{I}$.
/* laws */
  caused $context(i) = i_c$ **if** $sub(i_c, i)$. $\qquad$ (1)
  caused $context(i) = none$ **if** $\neg\bigvee_{i_c} sub(i_c, i)$. $\qquad$ (2)
  caused $sub_+(i_{cc}, i)$ **if** $sub(i_{cc}, i) \lor (sub(i_{cc}, i_c) \land sub_+(i_c, i))$. $\qquad$ (3)
  caused $\neg sub_+(i_{cc}, i)$ **if** $\neg\bigvee_{i_c}(sub(i_{cc}, i) \lor (sub(i_{cc}, i_c) \land sub_+(i_c, i)))$. $\qquad$ (4)

  constraint $state(i_c)=none \rightarrow context(i_c)=none \land \neg sub(i_c, i) \land$
  $\quad \neg member(i_c, a) \land \neg env(i_c, r)$. $\qquad$ (5)
  constraint $state(i) \neq none \rightarrow (i = top \leftrightarrow context(i) = none)$. $\qquad$ (6)
  constraint $\neg sub_+(i, i)$. $\qquad$ (7)
  caused $state(top)=open$. $\qquad$ (8)
  caused $state(i)=closed$ **if** $context(i) = i_c \land state(i_c) = closed$. $\qquad$ (9)

  $initiate(i, i_c)$ **causes** $state(i)=open \land sub(i_c, i)$. $\qquad$ (10)
  nonexecutable $initiate(i, i_c)$ **if** $state(i)=open$. $\qquad$ (11)
  default $\neg initiate(i, i_c)$. $\qquad$ (12)
  $finish(i)$ **causes** $state(i) = closed$. $\qquad$ (13)
  nonexecutable $finish(i)$ **if** $state(i)\neq open$. $\qquad$ (14)
  default $\neg finish(i)$. $\qquad$ (15)

Fig. 2. Specification $D_\mathcal{I}$ of the generic social interaction sort

Transitions. The initiation of interactions is modeled after the endogenous actions *initiate*. The execution of this action causes some interaction i to be opened within a specific context i_c (law 10). According to its only precondition (law 11), the interaction to be initiated can not be already open (which eliminates useless loops in the transition system). Thus, it could be either a non-existent or closed interaction. In this latter case, the execution of the initiate action causes its *reopening*. Note that the initiate action can not be executed if the specified interaction context is closed or non-existent in the *successor* state. These preconditions are defined indirectly through static laws 9 and 5, respectively. Note that this does not prevent some action to be executed if the specified context is not open in the current state – as long as the context is initiated as well in parallel. Thus, the specification allows to simultaneously initiate all the subinteractions of a given subtree in a single transition.

The *finish* action simply causes the interaction i to enter into the *closed* state (fluent dynamic law 13). The intended effect that every subinteraction is finished as well is not necessary to be directly specified, for the static law 9 provides an indirect means to achieve the same result. Thus, if a conference is prematurely closed, its program committee, submissions, reviewing teams, etc., are also automatically closed in a single transition. The only explicit precondition (causal law 14) establishes that non-existent interactions and already closed interactions can not be finished. This eliminates useless loops in the transition system. Last, note that, at this level of abstraction, it is not possible to identify the particular conditions which determine when some interaction must be initiated or finished. Therefore, their execution is disabled by default (action dynamic laws 12 and 15).

5   Agents

Agents shall be considered as a kind of role played by external software components within the computational society (specifically, within the *context* of a particular social interaction). Components attached to the society as agents will be regarded as autonomous (i.e. their state can be neither inspected nor altered by the social middleware[7]) and heterogeneous (they may be as complex as intelligent BDI components or simple as user interfaces). Any agent has a particular *purpose*, which represents the public goal that the software component purports to achieve as player of that role. In order to satisfy its purpose, an agent is empowered to *perform* different kinds of social actions, as will be described in the next section. Also, to account for complex scenarios, agents may structure their whole activity into a *role-playing* hierarchy of further agents. A *top-level* agent is one which is directly played by the software component, not by any other agent. The purposes of agents in a role-playing hierarchy somehow contributes to the satisfaction of the top-level agent's purpose.

For instance, the top-level agents illustrated in figure 1 are the *researchers* of the community. Software components running these agents are mostly plain user interfaces, since most of the tasks carried out by them are not amenable to automation. The activity of researchers within the community is decomposed into different roles. For instance, within the context of conferences, researchers may play the roles of *conference chairs*; within local committees, they may behave as *attendees*; within program committees, researchers may participate as *authors*, *PC chairs* and *PC members*; etc. The functionality of these roles may be even decomposed further. For instance, the activity of authors and PC chairs within the context of submissions is represented by the *submitter* and *submittee* roles, respectively; similarly, PC members behave as *reviewers* within the reviewing process (a role which could also be played directly by researchers of the community, to account for so-called *external* reviewers). Concerning agent purposes, a submitter agent purports to publish a particular scientific result in a given

---

[7] Unlike agents, software components attached to the society as resources, either to provide different computational services or, simply, to store information, are non-autonomous.

track of the conference (e.g. poster, oral, etc.)[8]; authors purport to publish their scientific results through the conference program; finally, researchers purport to increment its h-index. Clearly, the purposes of submitters are means to achieving the purpose of authors, which in turn contribute to the global purpose of researchers. Last, as can be observed in figure 1, it should be noted that agents may play several roles within the same context (e.g. a researcher playing both the role of author and PC member); and the same kind of role within different contexts (e.g. an author playing two submitter roles).

Similarly to social interactions, agents feature a run-time execution *state*, which indicates whether the agent is being *played*, i.e. is active within the society, or has been *abandoned*. Agents are played and abandoned by the middleware in response to different circumstances. For instance, a researcher agent is created by the society when a software component *enters* the community. Conversely, when the software component player of some researcher decides to *exit* the society, the corresponding role is abandoned. From that time on, that researcher is no longer considered a member of the community. Agents may also be created and destroyed by agents through the standard social actions *join* and *leave*, as described in the next section. For instance, authors are not created directly by software components, but by researchers when they *join* the conference. A role of this kind will be abandoned if a researcher decides to *leave* the conference prematurely. Last, agents may be played or destroyed automatically by the middleware when certain conditions hold. For instance, the submitter agent is automatically played for an author agent, as soon as that agent sets up the submission.

The structural and dynamic features described above are formally represented by the C+ specification D_A[9] , whose related signature and causal laws are shown in figure 3. The agent sort A subsumes any other particular, application-dependent agent sort, thereby formalising the common structure and dynamics of any agent. The specification also includes the declaration of the agent execution state sort S_A, and its accompanying objects *playing* and *abandoned*. The following paragraphs explain the major characteristics of the formalisation.

States. The major fluents that characterise the global state of some agent represent its run-time *state*, the social actions *performed* by the agent, and the *roles* in which its activity is decomposed. Besides these inertial fluents, the signature of the agent sort includes three statically determined fluents, defined by laws 16–21: the interaction *context* to which the agent belongs as a member; the *player* agent who plays that role (for non-top roles); and the transitive closure

---

[8] The purpose of some agent should not be confused with the *private goals* held by its software component (if implemented using a BDI language). The former represents the public goal that the component purports to achieve; the latter has no institutional significance, and may even clash with the public purpose (e.g. think of a researcher submitting a paper just to increment the number of submitted papers of the conference).

[9] Due to lack of space, the specification does not include the exogenous actions corresponding to the *enter* and *exit* external actions performed by software components.

of the *role* relationship. Last, the purpose of a given agent is represented by the application-dependent rules associated with the statically determined fluent *satisfied*.

:- **sorts**
   $\mathcal{A}$; $\mathcal{S}_\mathcal{A}$.
:- **objects**
   playing, abandoned :: $\mathcal{S}_\mathcal{A}$.
:- **constants**
   state($\mathcal{A}$) :: **inertialFluent**($\mathcal{S}_\mathcal{A}$+*none*);
   role($\mathcal{A},\mathcal{A}$), perform($\mathcal{A}$, Act) :: **inertialFluent**;
   context($\mathcal{A}$) :: **sdFluent**($\mathcal{I}$+*none*);
   player($\mathcal{A}$) :: **sdFluent**($\mathcal{A}$+*none*);
   role$_+$($\mathcal{A},\mathcal{A}$), satisfied($\mathcal{A}$) :: **sdFluent**;
   play($\mathcal{A},\mathcal{I}$), play($\mathcal{A},\mathcal{A},\mathcal{I}$), abandon($\mathcal{A}$) :: **action**.
:- **variables**
   a, a_p, a_{pp}:: $\mathcal{A}$.
/* laws */
   caused context(a) = i **if** member(i, a).   (16)
   caused context(a) = *none* **if** $\neg\bigvee_i$member(i, a).   (17)
   caused player(a) = a_p **if** role(a_p, a).   (18)
   caused player(a) = *none* **if** $\neg\bigvee_{a_p}$role(a_p, a).   (19)
   caused role$_+$(a_{pp}, a) **if** role(a_{pp}, a) $\vee$ (role(a_{pp}, a_p) $\wedge$ role$_+$(a_p, a)).   (20)
   caused $\neg$role$_+$(a_{pp}, a)
       if $\neg\bigvee_{a_p}$(role(a_{pp}, a) $\vee$ (role(a_{pp}, a_p) $\wedge$ role$_+$(a_p, a))).   (21)

   constraint state(a_p)=*none* $\rightarrow$ context(a_p)=*none* $\wedge$ player(a_p)=*none* $\wedge$
       $\neg$role(a_p, a) $\wedge$ $\neg$perform(a_p, α).   (22)
   constraint state(a) $\neq$ *none* $\rightarrow$ context(a) $\neq$ *none*.   (23)
   constraint $\neg$role$_+$(a, a).   (24)
   caused state(a)=*abandoned* **if** context(a) = i $\wedge$ state(i) = *closed*.   (25)
   caused state(a)=*abandoned* **if** player(a) = a_p $\wedge$ state(a_p) = *abandoned*. (26)
   default $\neg$satisfied(a).   (27)

   nonexecutable play(a, i) **if** state(a)=*playing*.   (28)
   default $\neg$play(a, i).   (29)
   play(a, i) **causes** member(i, a) $\wedge$ state(a) = *playing*.   (30)
   default $\neg$play(a, a_p, i).   (31)
   play(a, a_p, i) **causes** play(a, i).   (32)
   play(a, a_p, i) **causes** role(a_p, a).   (33)
   nonexecutable abandon(a) **if** state(a)$\neq$*playing*.   (34)
   default $\neg$abandon(a).   (35)
   abandon(a) **causes** state(a) = *abandoned*.   (36)

Fig 3. Specification $D_\mathcal{A}$ of the generic agent sort

Axioms 22–26 characterise the legal states of agents. Firstly, non-existent agent objects are undefined with respect to the major state parameters (static law 22). The next law establishes that any existent agent must belong to some interaction context. Causal law 24 together with the *player* fluent definition (laws 18–19) constrain the *role* graph to a forest of role-playing agents. Each tree of the forest represent the role-playing hierarchy associated with some top-level agent. Note that no causal law constrains in a particular way how role-playing hierarchies and the subinteraction tree of the society overlap, i.e. an agent may play some role in any interaction of the society. Note, however, that the run-time states of interactions and agents are constrained by static laws 25 and 26. They establish, in the form of a ramification effect, that some agent is abandoned if its interaction context is finished or its player is abandoned. Last, concerning purposes, since they are defined by specific subsorts, this generic specification simply establishes that the purpose is not satisfied by default (law 27).

Transitions. The signature of the generic agent sort $\mathcal{A}$ includes three actions related to the playing and abandonment of agents. Similarly to the *initiate* and *finish* actions, they are declared non-exogenous and are disabled by default. The first one, *play/2* (causal laws 28– 30), causes a given agent a to be played as member of a specific interaction context i. The only precondition explicitly declared prevents the execution of this action for agents that are already being played (so that the action allows for re-playing some abandoned agent). Axioms 5 and 25 indirectly prevent the execution of this action if the context is not open (i.e. non-existent or closed) in the successor state.

The role-playing hierarchy of agents is constructed through the *play/3* action (laws 31– 33), which causes some agent a to be played within a specific interaction context i by some player agent a_p. Only the last effect is declared explicitly (fluent dynamic law 33); the two first ones are indirectly caused through the action dynamic law 32. Moreover, this also causes the preconditions of *play/2* to be inherited.

The last action, *abandon* (laws 34– 36), causes the activity of some playing agent to be abandoned (maybe, without achieving its purpose). According to causal law 26, all its sub-roles will be also abandoned as an indirect effect. Thus, if a submission is prematurely finished when reviewers are in the middle of the revision process (e.g. because of a submitter withdrawal), then the submittee role will be automatically abandoned (according to static law 25) and the review process automatically finished (law 9); in turn, this will cause the reviewer roles to be abandoned as well.

6 Social Actions

Social actions provide agents with the means to change the state of the society in order to achieve their purposes. There are two major kinds of social actions: *communicative actions*, i.e. those actions performed in saying something, and *invocations* of resource operations, i.e. manipulating environmental resources. This paper just focuses on the first kind, and, particularly, on declarative speech acts. For instance, the *join* social action is an illocutionary act whereby an agent declares itself as the player of a new role within some interaction context.

Before delving into the discussion of these particular actions, however, the generic features shared by any kind of social action will be described.

Firstly, any institutional action is executed by the middleware infrastructure on behalf of a *performer* agent. In order for some social action to be successfully executed, *empowerments* and *permissions* of the prospective performer are also taken into account, i.e. common preconditions of general actions do not suffice, since agents are embedded in an institutional setting. Empowerments represent the institutional capabilities of agents, whereas permissions denote the circumstances that must be satisfied for exercising these capabilities. For instance, any researcher of the community is empowered to join a conference as author; however, this action is only permitted within the submission period, as defined by the program committee's calendar.

Social actions are brought about through the execution of *attempts*, viz. external actions executed – at will – by software components running some agent within the society. For instance, the attempt to make some researcher join a given conference is performed by a human user through the corresponding user interface. The attempts of causing unempowered agents to perform some action are simply ignored by the social middleware, i.e. the state of the society does not change at all. On the contrary, if the agent is empowered the execution of the attempt may alter the society in the following ways: a *forbidden* social action is created, if the agent is not permitted to perform the action in the current circumstances; or the action is *executed*, if the agent is indeed permitted to execute the action. The particular effects of the execution of some social action mostly depends on the particular kind of action.

The generic C+ sort *Act*, whose related signature and causal laws D_{Act} are shown in figure 4, represents any kind of social action. The specification D_{Act} also includes the sort S_{Act}, which represents the possible execution states of social actions: *executed* and *forbidden*.

States. The only inertial fluent of the specification represents the execution *state*, which initially holds the *none* value (representing that the action belongs to the pool of available objects). When the social action is attempted to be performed by an empowered agent it will take the *forbidden* or *executed* values, as formalised bellow. The *performer* agent is declared as a partial non-boolean fluent which is defined by causal laws 37 and 38. Only those actions whose execution was forbidden or were successfully executed are allowed to hold a performer (constraint 39).

Empowerments and permissions to perform a specified action are represented by the statically determined fluents *empowered* and *permitted*. Empowerment and permission rules heavily depends on the application domain, so that they will be specified by particular subsorts of interactions, agent, resources, and social actions. Nevertheless, the specification of these rules must guarantee that agents permitted to do some action are also empowered to do it (law 40). Last, empowerment and permissions for any kind of action and agent are false by default (laws 41 and 42).

```
:- sorts
      Act; S_Act.
:- objects
      forbidden, executed :: S_Act.
:- constants
```
state(Act) :: **inertialFluent**($\mathcal{S}_{Act}+none$);
performer(Act) :: **sdFluent**($\mathcal{A}+none$);
empowered(Act, $\mathcal{A}$), permitted(Act, $\mathcal{A}$) :: **sdFluent**;
attempt(Act, $\mathcal{A}$) :: **exogenousAction**;
execute(Act) :: **action**.
```
:- variables
      α:: Act.
/* laws */
```

caused performer(α) = a **if** perform(a, α).	(37)
caused performer(α) = $none$ **if** $\neg\bigvee_a$perform(a, α).	(38)

constraint state(α) $\neq none \leftrightarrow$ performer(α) $\neq none$.	(39)
constraint permitted(α, a) $\rightarrow$ empowered(α, a).	(40)
default $\neg$empowered(α, a).	(41)
default $\neg$permitted(α, a).	(42)

nonexecutable attempt(α, a) **if** state(α)$\neq none$ $\vee$ state(a)$\neq playing$.	(43)
attempt(α, a) **causes** perform(a, α) **if** empowered(α, a).	(44)
attempt(α, a) **causes** state(α)$=forbidden$ **if** empowered(α, a) $\wedge$ $\neg$permitted(α, a).	(45)
attempt(α, a) **causes** execute(α) **if** permitted(α, a).	(46)
default $\neg$execute(α).	(47)
execute(α) **causes** state(α)$=executed$.	(48)

Fig. 4. Specification D_{Act} of the generic social action sort

Transitions. The *attempts* of software components to make their agents perform some social action are represented as an *exogenous* C+ action; on the contrary, the *executions* of social actions by empowered and permitted agent are represented as common C+ actions. An attempt can not be executed if the action has already been processed or the intended performer is not currently being played (law 43). Otherwise, the following effects may be caused:

1. If the agent is empowered to perform the specified action, it is declared as its performer (law 44).
2. If the agent is empowered but it is not permitted to do that action, the forbidden attempt is registered accordingly (law 45).
3. If the agent is both empowered and permitted, the endogenous action *execute* is launched (action dynamic law 46). The only effect declared at this level of abstraction establishes that the action was indeed executed (fluent dynamic law 48). By default, the enactment of the *execute* action is disabled (action dynamic law 47).

Since all the previous effects are subject to the agent being empowered to do the action, if this condition does not obtain the state of the society does not change at all when the attempt is executed. Otherwise, if the agent is also permitted the execution of the attempt *counts as* executing the particular social action. Its particular effects and preconditions will be declared by the particular social action subsorts as described in the next subsection.

6.1 Standard Social Actions

Figure 1 shows several kinds of social actions of different levels of generality. On the one hand, we find social actions such as *submit, apply, assign* and *accept*. The first one allows authors to officially declare their intention to be evaluated; PC members express their reviewing preferences through the *apply* action; PC chairs are empowered to *assign* PC members to reviewing processes; last, when behaving as submittees, PC chairs are also empowered to *accept* some submitted paper as part of the conference program. These actions, though amenable to be applied within different application domains, are clearly related to particular types of social interactions (submissions and task assignments). On the other hand, social actions such as *set up* and *join* are completely independent of any application domain and social interaction type. Hence, they will be considered as part of a catalogue of *standard* social actions. This catalogue includes a number of declarative speech acts, which allow agents to manipulate the interaction space of the society as well as the role-playing hierarchies. In particular, the catalogue includes the *set up* and *close* actions, to initiate and finish interactions; the *join* action, which allows some agent to declare that it plays some role within some context; and the *leave* and *fire* actions, which cause some agent to be abandoned (a role played by the performer agent in the former case, and some arbitrary agent in the latter). Figure 5 shows the specification D_{join} of the *join* action, in order to illustrate the pattern of specification of social actions. This specification is part of a larger action description $D_{StandardLib}$ which includes the specification of the other standard actions as well.

Firstly, the new sort *join* is declared as a subsort of the generic social action sort *Act*. Then, the super-sort specification is extended with new inertial fluents. In this case, the attributes *new* and *context*, which identify the new role to be played and its desired context. Last, the particular effects of the social action are declared through causal laws which extends the predefined effect of the *execute* action (see figure 3). In particular, the execution of a join action causes the enactment of the internal action *play/3* (causal law 49). In a similar way, the *close, set up* and *leave/fire* actions are counterparts of the *finish, initiate* and *abandon* internal actions, respectively. As a consequence of the action dynamic law 49, the explicit preconditions of the internal action *play/3* are inherited by the social action *join*. Thus, even if some agent is empowered and permitted to execute a given action, the execution will be prevented if it is not well-formed. Finally, the specification of a particular kind of social action may declare additional preconditions and/or constraints to its empowerment or permission rules (e.g. the only agent which is empowered to *leave* a given role is its player).

```
:- sorts
    Act ≫ join.
:- constants
    new(join) :: inertialFluent(𝒜);
    context(join) :: inertialFluent(ℐ).
:- variables
    α_join :: join.
/* laws */
    execute(α_join) ∧ attempt(α_join, a_p) causes play(a, a_p, i).          (49)
```

Fig. 5. Specification D_{join} of the *join* social action

7   Example

This section illustrates the dynamics of the social middleware with a small example in which several attempts are processed. In particular, figure 6 describes a small part of the transition system representing the dynamics of a simple research community. This transition system is described by the action description D_{ResCom}, which is made up of the action descriptions D_{Speech}, $D_{StandardLib}$, D_{Conf} and D_{Domain}. This last component includes the declaration of two researchers r_1 and r_2, a conference c_1 and its program committee pc_1, an author a_1, a PC member pcm_1, and two *join* actions α_1 and α_2. Two states and four transitions are shown in figure 6. In the first state, s_1, the two researchers are in a playing state, the conference is open for submissions and the social actions belong to the pool of available objects; the first one, α_1, would allow the performer agent to join the program committee as author a_1; the second one, α_2, as PC member (agents a_1 and pcm_1 belong to the pool of non-existent objects). Let us suppose that both agents, r_1 and r_2 are empowered and permitted to perform action α_1, and that none of them is empowered to perform α_2[10].

Leading from state s_1, several transitions may be considered in relation to the exogenous attempts to execute actions α_1 and α_2. Firstly, it may happen that components attempt their agents to perform neither α_1 nor α_2. This possibility is represented by the anonymous transition t_1. Since no actions apply, inertia is in effect and the transition results in a loop. Secondly, it may happen that some component attempts its agent to perform action α_1[11]. For instance, transition t_2 illustrates this case when the performer is researcher r_1. Since this agent is empowered and permitted to do this action, the attempt causes the execution of the *join* declaration and, eventually, the playing of the specified author role, as described by the resulting state s_2. Thirdly, it may be the case that only action

---

[10] The rules for establishing these conclusions are part of the specifications D_{author} and $D_{PCMember}$ of the application-dependent sorts *author* and *PC member* (which are part of the specification D_{Conf} and are not shown here due to lack of space).

[11] It should be noted that no action can be attempted to be performed by more than one empowered agent, since the value of the *performer* fluent would not be unique in that case.

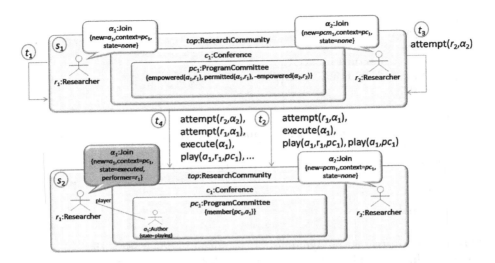

Fig. 6. Attempt processing in the conference management scenario

α_2 is attempted to be executed by some researcher. In these scenarios, since the researchers are not empowered to perform this action, the corresponding transitions do not cause any institutional change, i.e. they are loops. Transition t_3 of figure 6 illustrates this possibility when the performer agent is r_2. Last, it may happen that both actions are attempted. In these cases, the effects are those obtained through the execution of action α_1, since α_2 does not contribute with any change in the society. Thus, transition t_4 has the same resulting state as transition t_2.

8   Discussion

This paper has presented a programming language approach to the implementation of social processes as computational societies. This approach differs from other proposals in the literature in several respects. Firstly, there is an attempt to capture the computational aspects of societal issues through a dedicated programming language. On the contrary, other approaches (e.g. [7]) proceed by extending BDI programming languages with organizational constructs. As a consequence, the semantics of the organizational concepts has strong dependencies with the agent architecture of choice, which limits their applicability to radically different architectures (e.g. a simple Java component). Thus, we endorse a strict separation of concerns between interaction (societal issues) and computation (software components). In this regard, the proposal of this paper is more closely aligned with the attempt at designing a programming language for organizational artifacts reported in [6]. In contrast with this work, however, we demand neither a reference to agent components nor its programs from the side of the social middleware. Hence, the autonomy of agent components is respected altogether. For instance, this works in the interest of a greater deployment flexibility (e.g. some agent role being alternatively played by components at different locations).

Secondly, we endorse a strong support for modularity and reusability. With respect to the former, the language offers four programming abstractions (social interactions, agents, resources and social actions) and two decomposition strategies (role-playing agent and sub-interaction hierarchies) which allow to partition the whole set of rules of a computational society into manageable modules. As far as reusability is concerned, the implementation of application-dependent sorts (e.g. paper submissions, reviewing process) do not start from scratch. On the contrary, their implementation builds upon the common structure and dynamics encoded in the generic sorts $\mathcal{I}$, $\mathcal{A}$, etc. Moreover, the subsort mechanism can also be exploited in the design of application-independent libraries, such as the *standard* library of social actions described in section 6.1. Concerning these matters, other approaches to the problem of producing high-level, executable specifications of computational societies such as [2,5,18] fall short of providing adequate social structuring and reusability mechanisms. Moreover, they mostly concentrate on the semantics of normative positions, leaving aside key concepts such as social actions and environmental resources. In this regard, the proposed model of social interaction smoothly integrates the major communicative, environmental and organisational dimensions of multiagent systems.

The CCALC implementation of the run-time semantics of the language and a simplified version of the rules of conference management, together with some prediction and planning queries, can be found online[12]. This implementation has primarily served to validate the formal specification of the run-time semantics of the language. In this regard, the ability to represent indirect effects and preconditions, ramifications, etc., has shown C+ as a very convenient formalism. However, the inefficiencies of the grounding mechanism and completion process make the applicability of CCALC in a real setting quite debatable.

The presented run-time semantics is intended as the first step towards a pure societal programming language. Current work aims at the other major component of the language: its type system. Also, we aim at addressing the limitations of the social ontology by accommodating computational resources, commitments [8] and full-fledged communicative actions [4]. Concerning formal matters, the current approach to modularity based upon the CCALC include directive is rather limited. In this regard, our approach may greatly benefit from [13,9]. Last, we are working on a web-based, RESTful implementation of the language [16].

References

1. Akman, V., Erdogan, S.T., Lee, J., Lifschitz, V., Turner, H.: Representing the zoo world and the traffic world in the language of the causal calculator. Artif. Intell 153(1-2), 105–140 (2004)

---

[12] http://zenon.etsii.urjc.es/~jserrano/speech/apps/c+apps.tgz

2. Artikis, A., Sergot, M., Pitt, J.: Specifying norm-governed computational societies. ACM Transactions on Computational Logic 10(1) (2009)
3. Baral, C., Gelfond, M.: Logic programming and reasoning about actions. In: Handbook of Temporal Reasoning in Artificial Intelligence, ch. 13, pp. 389–496. Elsevier, Amsterdam (2005)
4. Boella, G., Damiano, R., Hulstijn, J., van der Torre, L.: A common ontology of agent communication languages: Modeling mental attitudes and social commitments using roles. Applied Ontology 2(3-4), 217–265 (2007)
5. Cliffe, O., De Vos, M., Padget, J.A.: Answer set programming for representing and reasoning about virtual institutions. In: Inoue, K., Satoh, K., Toni, F. (eds.) CLIMA 2006. LNCS, vol. 4371, pp. 60–79. Springer, Heidelberg (2007)
6. Dastani, M., Tinnemeier, N., Meyer, J.-J.C.: A programming language for normative multi-agent systems. In: Dignum, V. (ed.): Multi-Agent Systems: Semantics and Dynamics of Organizational Models. IGI Global (in press)
7. Dennis, L.A., Fisher, M., Hepple, A.: Language constructs for multi-agent programming. In: Sadri, F., Satoh, K. (eds.) CLIMA VIII. LNCS, vol. 5056, pp. 137–156. Springer, Heidelberg (2008)
8. Desai, N., Chopra, A.K., Singh, M.P.: Representing and reasoning about commitments in business processes. In: AAAI XXII, pp. 1328–1333. AAAI Press, Menlo Park (2007)
9. Desai, N., Singh, M.P.: A modular action description language for protocol composition. In: AAAI XXII, pp. 962–967. AAAI Press, Menlo Park (2007)
10. Esteva, M., Rosell, B., Rodríguez-Aguilar, J.A., Arcos, J.L.: Ameli: An agent-based middleware for electronic institutions. In: AAMAS III, pp. 236–243. IEEE Computer Society Press, Los Alamitos (2004)
11. Giunchiglia, E., Lee, J., Lifschitz, V., McCain, N., Turner, H.: Nonmonotonic causal theories. Artif. Intell. 153(1-2), 49–104 (2004)
12. Hübner, J.F., Sichman, J.S., Boissier, O.: S-moise+: A middleware for developing organised multi-agent systems. In: Boissier, O., Dignum, V., Matson, E., Sichman, J.S. (eds.) COIN 2006. LNCS, vol. 3913, pp. 64–78. Springer, Heidelberg (2006)
13. Lifschitz, V., Ren, W.: A modular action description language. In: AAAI XXI. AAAI Press, Menlo Park (2006)
14. Mehta, N.R., Medvidovic, N., Phadke, S.: Towards a taxonomy of software connectors. In: ICSE XXII, pp. 178–187. ACM Press, New York (2000)
15. Reiter, R.: On closed world databases. In: Gallaire, Minker (eds.) Logic and Databases, pp. 55–76. Plenum Press, New York (1978)
16. Saugar, S., Serrano, J.M.: A web-based virtual machine for developing computational societies. In: Klusch, M., Pěchouček, M., Polleres, A. (eds.) CIA 2008. LNCS, vol. 5180, pp. 162–176. Springer, Heidelberg (2008)
17. Serrano, J.M., Saugar, S.: Operational semantics of multiagent interactions. In: Durfee, E.H., Yokoo, M., Huhns, M.N., Shehory, O. (eds.) AAMAS VI, pp. 889–896. IFAAMAS (2007)
18. Viganó, F., Colombetti, M.: Specification and verification of institutions through status functions. In: Noriega, P., Vázquez-Salceda, J., Boella, G., Boissier, O., Dignum, V., Fornara, N., Matson, E. (eds.) COIN 2006. LNCS, vol. 4386, pp. 115–129. Springer, Heidelberg (2007)

Embedding Linear-Time Temporal Logic into Infinitary Logic: Application to Cut-Elimination for Multi-agent Infinitary Epistemic Linear-Time Temporal Logic

Norihiro Kamide

Waseda Institute for Advanced Study
1-6-1 Nishi Waseda, Shinjuku-ku, Tokyo 169-8050, Japan
logician-kamide@aoni.waseda.jp

Abstract. Linear-time temporal logic (LTL) is known as one of the most useful logics for verifying concurrent systems, and infinitary logic (IL) is known as an important logic for formalizing common knowledge reasoning. The research fields of both LTL and IL have independently been developed each other, and the relationship between them has not yet been discussed before. In this paper, the relationship between LTL and IL is clarified by showing an embedding of LTL into IL. This embedding shows that globally and eventually operators in LTL can respectively be represented by infinitary conjunction and infinitary disjunction in IL. The embedding is investigated by two ways: one is a syntactical way, which is based on Gentzen-type sequent calculi, and the other is a semantical way, which is based on Kripke semantics. The cut-elimination theorems for (some sequent calculi for) LTL, an infinitary linear-time temporal logic ILT_ω (i.e., an integration of LTL and IL), a multi-agent infinitary epistemic linear-time temporal logic $IELT_\omega$ and a multi-agent epistemic bounded linear-time temporal logic ELT_l are obtained as applications of the resulting embedding theorem and its extensions and modifications. In particular, the cut-elimination theorem for $IELT_\omega$ gives a new proof-theoretical basis for extremely expressive time-dependent multi-agent logical systems with common knowledge reasoning.

1 Introduction

1.1 Linear-Time Temporal Logic and Infinitary Logic

Linear-time temporal logic (LTL), which has the temporal operators X (next), G (globally) and F (eventually), has widely been studied in order to verify and specify concurrent systems [4,5,15,24,31]. Gentzen-type sequent calculi for LTL and its neighbors have also been introduced by many researchers [1,13,22,23,25,26]. A sequent calculus LT_ω for (until-free) LTL was introduced by Kawai, and the cut-elimination and completeness theorems for this calculus were proved [13]. A *2-sequent calculus* $2S\omega$ for (until-free) LTL, which is a natural extension of the

M. Fisher, F. Sadri, and M. Thielscher (Eds.): CLIMA IX, LNAI 5405, pp. 57–76, 2009.
© Springer-Verlag Berlin Heidelberg 2009

usual sequent calculus, was introduced by Baratella and Masini, and the cut-elimination and completeness theorems for this calculus were proved based on an analogy between LTL and Peano arithmetic with ω-rule [1]. A direct syntactical equivalence between Kawai's LT_ω and Baratella-Masini's $2\mathrm{S}\omega$ was shown by introducing the translation functions that preserve cut-free proofs of these calculi [9]. In the present paper, LT_ω is used to investigate an embedding of LTL into infinitary logic, since LT_ω gives a simple proof of the syntactical embedding theorem discussed.

Infinitary logic (IL), which has the connectives of infinitary conjunction $\bigwedge$ and infinitary disjunction $\bigvee$, has been studied by many logicians [6,19,21,27]. A historical survey of the proof-theoretic use of the infinitary connectives was presented in [6]. A survey of Barwise's works and their developments on IL and admissible sets was presented in [14]. Such works include to show how infinitary modal logic ties non-well-founded set theory with the infinitary first-order logic of bisimulation. It is hoped that IL is useful in computer science, since the usual regular programs including Kleene iteration and many further natural programming constructs are infinitary as well. In [2], the modal bisimulation invariance and safety theorems, which link modal logic with computational process theories, were shown for infinitary modal logic. Gentzen-type sequent calculi for modal extensions of IL have also been studied from the point of view of common knowledge reasoning and game theory [11,12,28].

Although the research areas of both LTL and IL have independently been evolved as long as more than 30 years, the connection between them has not yet been found before. One of the aims of this paper is thus to clarify the connection between them giving a faithful embedding of LTL into IL. This embedding shows that G and F in LTL can respectively be represented by (countable) $\bigwedge$ and $\bigvee$ in IL. The embedding is investigated in both a syntactical way, which is based on cut-free Gentzen-type sequent calculi, and a semantical way, which is based on Kripke semantics.

1.2 Proposed Embedding Theorems

The *syntactical embedding theorem* presented is based on Gentzen-type sequent calculi LT_ω (for LTL) and LK_ω (for IL), and an alternative proof of the cut-elimination theorem for LT_ω can be obtained using this embedding theorem. The syntactical embedding theorem provides the fact "a sequent $\Rightarrow \alpha$ is provable in LT_ω if and only if the sequent $\Rightarrow f(\alpha)$ is provable in LK_ω, where f is a certain embedding function." The essential idea of the embedding function f is to represent the following informal intepretations: for any propositional variable p of LTL, $f(p) = p$, $f(\mathrm{X}^i p) = p_i$, $f(\mathrm{X}^i \mathrm{G}\alpha) = \bigwedge\{f(\mathrm{X}^{i+j}\alpha) \mid j \in \omega\}$ and $f(\mathrm{X}^i \mathrm{F}\alpha) = \bigvee\{f(\mathrm{X}^{i+j}\alpha) \mid j \in \omega\}$. The syntactical embedding theorem can also be adapted to Baratella and Masini's $2\mathrm{S}\omega$ by using the translation proposed in [9].

Although the syntactical embedding theorem gives a proof-theoretical interpretation of the connection between LTL and IL, a semantical interpretation of the same connection cannot be obtained directly. On the other hand, LTL is usually defined by a semantical way. Indeed, the model checking methods using

LTL are based on a purely semantical expression. Moreover, most of the LTL users are not familiar with the Gentzen-type proof theory. Thus, a semantical interpretation of the connection between LTL and IL is needed. In order to obtain such a semantical interpretation, a semantical version of the syntactical embedding theorem, called the *semantical embedding theorem*, is presented.

The importance of this work may be that the methodologies which have independently been developed in the research fields of both LTL and IL can be connected. In our opinion, the theoretical understanding of IL is superior to that of LTL, since there are a number of traditional studies concerning IL. On the other hand, the practical usefulness of LTL is superior to that of IL, since there are a number of computer science applications using LTL. We hope that both the theoretical advantages of IL and the practical advantages of LTL are combined and integrated using the idea of the proposed embedding theorems.

In this respect, the cut-elimination theorems for an infinitary extension ILT_ω of LT_ω (i.e., an integration of LT_ω and LK_ω), a first-order multi-agent epistemic extension IELT_ω of ILT_ω and a modified multi-agent subsystem ELT_l of IELT_ω are obtained using some extensions and modifications of the syntactical embedding theorem.

1.3   Cut-Elimination for Multi-agent Sequent System

The cut-elimination theorem for IELT_ω gives a reasonable proof-theoretical basis for extremely expressive time-dependent multi-agent logical systems with common knowledge reasoning, since the cut-elimination theorem guarantees the consistency of IELT_ω, i.e., the unprovability of the empty sequent "$\Rightarrow$", and the conservativity over any fragments of IELT_ω, i.e., IELT_ω is recognized as a conservative extension of LT_ω, LK_ω and ILT_ω. Moreover, a modified propositional subsystem ELT_l of IELT_ω, which cannot express the notion of common knowledge, is shown to be decidable by using a modified syntactical embedding theorem.

The advantage of the proposed embedding-based cut-elimination proof is that the proof is very easy and simple than other proposals. Indeed, it is known that the direct Gentzen-style cut-elimination proof of LK_ω is very difficult and complex (see e.g., [12,27,28]). Thus, it is very difficult to obtain the direct Gentzen-style proofs of the cut-elimination theorems for ILT_ω and IELT_ω. Such a difficulty can be avoided by our embedding method.

The advantage of IELT_ω may be that it is very expressive enough to deal with multi-agent (by the S4-type multi-modal epistemic operators) reasoning, time-dependent (by the temporal operators in LTL) reasoning and common knowledge (by the infinitary connectives) reasoning. IELT_ω is constructed based on a sequent calculus S4_ω for a multi-agent infinitary epistemic logic [12,28,29], i.e., IELT_ω is an integration of S4_ω and LT_ω.

Some variants of S4_ω have been studied by some researchers (e.g., [11,12,29]). For example, an application to game theory using a KD4-version of S4_ω was proposed by Kaneko and Nagashima [12]. In [12], to express the common knowledge concept explicitly, the infinitary connectives were used effectively. This logic was called by them *game logic* GL. An embedding of a *common knowledge logic* CKL into GL was shown by Kaneko [11], where CKL is an epistemic logic with one knowledge operator for each player and a common knowledge operator [7,18].

The present paper's result for IELT$_\omega$ is intended to give a proof-theoretical basis for extended and combined reasoning with common knowledge and time. As far as we know, there is no study for combining a common knowledge logic with time, although there are a lot of studies dealing with both knowledge and time (see e.g., [3,8,20,30]). Combining (or cooperating) common knowledge with time may be useful for some computer science applications such as security issues [3], since a kind of secret key is regarded as the common knowledge of multi-agents. We hope that IELT$_\omega$ is useful for expressing various time-dependent multi-agent systems with common knowledge, although the result of this paper is purely proof-theoretical. We also hope that our embedding method is useful for showing the cut-elimination theorems for a wide range of temporal and infinitary multi-agent logics. Indeed, our method is also applicable to the cut-elimination theorems for the K4-type, KT-type and K-type modifications of IELT$_\omega$. Our method is also applicable to the cut-elimination theorems for some intuitionistic versions of some fragments of IELT$_\omega$ and its modifications, although such a result is not addressed in this paper.

1.4   Combining Common Knowledge with Time

The notion of common knowledge was probably first introduced by Lewis [16]. This notion is briefly explained below. Let I be a fixed set of agents and α be an idea. Suppose α belongs to the common knowledge of I, and i and j are some members of I. Then, we have the facts "both i and j know α", "i knows that j knows α" and "j knows that i knows α". Moreover, we also have the facts "i knows that j knows that i knows α", and vice versa. Then, these nesting structures develop an infinite hierarchy as a result.

We try to express such a infinite nesting based on IELT$_\omega$. IELT$_\omega$ has the knowledge operators $K_1, K_2, ..., K_n$, in which a formula $K_i \alpha$ means "the agent i knows α". The common knowledge of a formula α is defined below [12]. For any $m \geq 0$, an expression $K(m)$ means the set

$\{K_{i_1} K_{i_2} \cdots K_{i_m} \mid$ *each* K_{i_t} *is one of* $K_1, ..., K_n$ *and* $i_t \neq i_{t+1}$ *for all* $t = 1, ..., m - 1\}$.

When $m = 0$, $K_{i_1} K_{i_2} \cdots K_{i_m}$ is interpreted as the null symbol. The common knowledge $C\alpha$ of α is defined by using the infinitary connective $\bigwedge$ as:

$$C\alpha := \bigwedge \{K\alpha \mid K \in \bigcup_{m \in \omega} K(m)\}.$$

Then, the formula $C\alpha$ means "α is a common knowledge of agents".

Various kinds of time-dependent situations with common knowledge can be expressed using IELT$_\omega$. For example, a liveness property "If the login password of a computer is regarded as the common knowledge, then we will eventually be able to login the computer" can be expressed as

$$G(C \text{ } password \rightarrow F \text{ } login).$$

1.5   Summary of This Paper

The contents of this paper are summarized as follows.

In Section 2, firstly, the sequent calculi LT_ω and LK_ω are introduced for LTL and IL, respectively, secondly, the embedding function from the language of LT_ω into that of LK_ω is presented, and finally the syntactical embedding theorem is proved based on this embedding function.

In Section 3, the semantics of LTL and IL are introduced, and the semantical embedding theorem is shown.

In Section 4, the cut-elimination theorems for ILT_ω, $IELT_\omega$ and ELT_l, which are the main results of this paper, are shown by using some extensions and modifications of the syntactical embedding theorem. In addition, ELT_l is shown to be decidable. The cut-elimination theorems for ILT_ω and $IELT_\omega$ may not be obtained by using other proof methods because of the complexity. This is the reason why we use the simple and easy method based on the embedding theorems.

In Section 5, this paper is concluded, and some remarks on the K4-type, KT-type and K-type variants of $IELT_\omega$ are addressed.

2   Syntactical Embedding

2.1 LT_ω for LTL

Formulas of the (until-free) propositional linear-time temporal logic (LTL) are constructed from countable propositional variables, $\rightarrow$ (implication), $\wedge$ (conjunction), $\vee$ (disjunction), $\neg$ (negation), G (globally), F (eventually) and X (next). Lower-case letters $p, q, \ldots$ are used to denote propositional variables, Greek lower-case letters $\alpha, \beta, \ldots$ are used to denote formulas, and Greek capital letters $\Gamma, \Delta, \ldots$ are used to represent finite (possibly empty) sets of formulas. For any $\sharp \in \{G, F, X\}$, an expression $\sharp \Gamma$ is used to denote the set $\{\sharp \gamma \mid \gamma \in \Gamma\}$. The symbol $\equiv$ is used to denote the equality of sets of symbols. The symbol ω is used to represent the set of natural numbers. An expression $X^i \alpha$ for any $i \in \omega$ is used to denote $\overbrace{XX \cdots X}^{i} \alpha$, e.g., $(X^0 \alpha \equiv \alpha)$, $(X^1 \alpha \equiv X\alpha)$ and $(X^{n+1} \alpha \equiv X^n X\alpha)$. Lower-case letters i, j and k are used to denote any natural numbers. An expression of the form $\Gamma \Rightarrow \Delta$ is called a *sequent*. An expression $L \vdash S$ is used to denote the fact that a sequent S is provable in a sequent calculus L.

Kawai's LT_ω for LTL is then presented below.

Definition 1 (LT_ω). *The initial sequents of LT_ω are of the form: for any atomic formula p,*

$$X^i p \Rightarrow X^i p.$$

The structural rules of LT_ω are of the form:

$$\frac{\Gamma \Rightarrow \Delta, X^i \alpha \quad X^i \alpha, \Sigma \Rightarrow \Pi}{\Gamma, \Sigma \Rightarrow \Delta, \Pi} \text{ (cut)}$$

$$\frac{\Gamma \Rightarrow \Delta}{X^i\alpha, \Gamma \Rightarrow \Delta} \text{ (we-left)} \qquad \frac{\Gamma \Rightarrow \Delta}{\Gamma \Rightarrow \Delta, X^i\alpha} \text{ (we-right).}$$

The logical inference rules of LT_ω *are of the form:*

$$\frac{\Gamma \Rightarrow \Sigma, X^i\alpha \quad X^i\beta, \Delta \Rightarrow \Pi}{X^i(\alpha \rightarrow \beta), \Gamma, \Delta \Rightarrow \Sigma, \Pi} \text{ (}\rightarrow\text{left)} \qquad \frac{X^i\alpha, \Gamma \Rightarrow \Delta, X^i\beta}{\Gamma \Rightarrow \Delta, X^i(\alpha \rightarrow \beta)} \text{ (}\rightarrow\text{right)}$$

$$\frac{X^i\alpha, \Gamma \Rightarrow \Delta}{X^i(\alpha \wedge \beta), \Gamma \Rightarrow \Delta} \text{ (}\wedge\text{left1)} \qquad \frac{X^i\beta, \Gamma \Rightarrow \Delta}{X^i(\alpha \wedge \beta), \Gamma \Rightarrow \Delta} \text{ (}\wedge\text{left2)}$$

$$\frac{\Gamma \Rightarrow \Delta, X^i\alpha \quad \Gamma \Rightarrow \Delta, X^i\beta}{\Gamma \Rightarrow \Delta, X^i(\alpha \wedge \beta)} \text{ (}\wedge\text{right)} \qquad \frac{X^i\alpha, \Gamma \Rightarrow \Delta \quad X^i\beta, \Gamma \Rightarrow \Delta}{X^i(\alpha \vee \beta), \Gamma \Rightarrow \Delta} \text{ (}\vee\text{left)}$$

$$\frac{\Gamma \Rightarrow \Delta, X^i\alpha}{\Gamma \Rightarrow \Delta, X^i(\alpha \vee \beta)} \text{ (}\vee\text{right1)} \qquad \frac{\Gamma \Rightarrow \Delta, X^i\beta}{\Gamma \Rightarrow \Delta, X^i(\alpha \vee \beta)} \text{ (}\vee\text{right2)}$$

$$\frac{\Gamma \Rightarrow \Delta, X^i\alpha}{X^i\neg\alpha, \Gamma \Rightarrow \Delta} \text{ (}\neg\text{left)} \qquad \frac{X^i\alpha, \Gamma \Rightarrow \Delta}{\Gamma \Rightarrow \Delta, X^i\neg\alpha} \text{ (}\neg\text{right)}$$

$$\frac{X^{i+k}\alpha, \Gamma \Rightarrow \Delta}{X^iG\alpha, \Gamma \Rightarrow \Delta} \text{ (Gleft)} \qquad \frac{\{\Gamma \Rightarrow \Delta, X^{i+j}\alpha\}_{j\in\omega}}{\Gamma \Rightarrow \Delta, X^iG\alpha} \text{ (Gright)}$$

$$\frac{\{X^{i+j}\alpha, \Gamma \Rightarrow \Delta\}_{j\in\omega}}{X^iF\alpha, \Gamma \Rightarrow \Delta} \text{ (Fleft)} \qquad \frac{\Gamma \Rightarrow \Delta, X^{i+k}\alpha}{\Gamma \Rightarrow \Delta, X^iF\alpha} \text{ (Fright).}$$

It is remarked that (Gright) and (Fleft) have infinite premises. The sequents of the form: $X^i\alpha \Rightarrow X^i\alpha$ for any formula α are provable in LT_ω. This fact can be proved by induction on the complexity of α. The original Kawai's sequent calculus [13] is, strictly speaking, obtained from the above defined LT_ω by replacing the initial sequents and structural rules by the initial sequents and structural rules only with $i = k = 0$, i.e., every X^i is deleted from the forms of the initial sequents and structural rules. Obviously, this restriction does not change the the the provability of sequents, i.e., it is not a restriction. It is also remarked that in [13], the next-time operator is not used as a modal operator but used as a special symbol. The cut-elimination theorem for LT_ω was shown by Kawai in [13]. An alternative proof of this theorem will be given in Section 4.1.

2.2 LK$_\omega$ for IL

Formulas of the propositional classical infinitary logic (IL) are constructed from (countable) propositional variables, $\rightarrow$ (implication), $\neg$ (negation), $\bigwedge$ (infinitary conjunction) and $\bigvee$ (infinitary disjunction). For $\bigwedge$ and $\bigvee$, if Φ is a countable non-empty set of formulas of IL, then $\bigwedge\Phi$ and $\bigvee\Phi$ are also formulas of IL. It is also remarked that $\bigwedge\{\alpha\}$ and $\bigvee\{\alpha\}$ are equivalent to α. It is noted that the standard binary connectives $\wedge$ (conjunction) and $\vee$ (disjunction) are regarded as special cases of $\bigwedge$ and $\bigvee$, respectively. It is intended in this section that $\bigwedge$ and $\bigvee$ are a countable infinitary conjunction and a countable infinitary disjunction, respectively. For the sake of brevity and compatibility, the precise definition of the (set of) formulas of IL is not addressed here. Such a definition is given for an extended language discussed in Section 4.2.

A sequent calculus LK_ω for IL is then presented below.

Definition 2 (LK_ω). *The initial sequents of* LK_ω *are of the form: for any atomic formula* p,

$$p \Rightarrow p.$$

The structural rules of LK_ω *are of the form:*

$$\frac{\Gamma \Rightarrow \Delta, \alpha \quad \alpha, \Sigma \Rightarrow \Pi}{\Gamma, \Sigma \Rightarrow \Delta, \Pi} \ (\mathrm{cut}^0)$$

$$\frac{\Gamma \Rightarrow \Delta}{\alpha, \Gamma \Rightarrow \Delta} \ (\text{we-left}^0) \qquad \frac{\Gamma \Rightarrow \Delta}{\Gamma \Rightarrow \Delta, \alpha} \ (\text{we-right}^0).$$

The logical inference rules of LK_ω *are of the form:*

$$\frac{\Gamma \Rightarrow \Sigma, \alpha \quad \beta, \Delta \Rightarrow \Pi}{\alpha \rightarrow \beta, \Gamma, \Delta \Rightarrow \Sigma, \Pi} \ (\rightarrow\text{left}^0) \qquad \frac{\alpha, \Gamma \Rightarrow \Delta, \beta}{\Gamma \Rightarrow \Delta, \alpha \rightarrow \beta} \ (\rightarrow\text{right}^0)$$

$$\frac{\Gamma \Rightarrow \Delta, \alpha}{\neg\alpha, \Gamma \Rightarrow \Delta} \ (\neg\text{left}^0) \qquad \frac{\alpha, \Gamma \Rightarrow \Delta}{\Gamma \Rightarrow \Delta, \neg\alpha} \ (\neg\text{right}^0)$$

$$\frac{\alpha, \Gamma \Rightarrow \Delta \ (\alpha \in \Theta)}{\bigwedge\Theta, \Gamma \Rightarrow \Delta} \ (\bigwedge\text{left}) \qquad \frac{\{\ \Gamma \Rightarrow \Delta, \alpha \ \}_{\alpha \in \Theta}}{\Gamma \Rightarrow \Delta, \bigwedge\Theta} \ (\bigwedge\text{right})$$

$$\frac{\{\ \alpha, \Gamma \Rightarrow \Delta \ \}_{\alpha \in \Theta}}{\bigvee\Theta, \Gamma \Rightarrow \Delta} \ (\bigvee\text{left}) \qquad \frac{\Gamma \Rightarrow \Delta, \alpha \ (\alpha \in \Theta)}{\Gamma \Rightarrow \Delta, \bigvee\Theta} \ (\bigvee\text{right})$$

where Θ *denotes a non-empty countable set of formulas.*

The "0" in the rule names in LK_ω means that these rules are the special cases of the corresponding rules of LT_ω, i.e., the case that i of X^i is 0.

It is remarked that instead of the rules (we-left^0) and (we-right^0), the following single weakening rule was used in [12,28]:

$$\frac{\Gamma \Rightarrow \Delta}{\Pi, \Gamma \Rightarrow \Delta, \Sigma} \ (\text{we}).$$

The cut-free equivalence between LK_ω and the modified system by replacing (we-left^0) and (we-right^0) by (we) can be shown, and hence the same cut-elimination result as in the modified LK_ω with (we) can also be obtained.

The following cut-elimination theorem is known (see e.g., [6,12,28]).

Theorem 3. *The rule* (cut^0) *is admissible in cut-free* LK_ω.

2.3   Embedding from LTL into IL

Definition 4 (Embedding). *We fix a countable non-empty set Φ of atomic formulas, used as a component of the language of* LTL *(or* LT_ω*), and define the sets $\Phi_i := \{p_i \mid p \in \Phi\}$ $(1 \le i \in \omega)$ and $\Phi_0 := \Phi$ of atomic formulas. The language $\mathcal{L}_{LTL}$ of* LTL *(or* LT_ω*) is defined by using Φ, $\to, \wedge, \vee, \neg$, X, G and F. The language $\mathcal{L}_{IL}$ of* IL *(or* LK_ω*) is defined by using $\bigcup_{i \in \omega} \Phi_i$, $\to, \neg$, $\bigwedge$ and $\bigvee$. The binary versions of $\bigwedge$ and $\bigvee$ are also denoted as $\wedge$ and $\vee$, respectively, and these binary symbols are included in the definition of $\mathcal{L}_{IL}$.*

A mapping f from $\mathcal{L}_{LTL}$ to $\mathcal{L}_{IL}$ is defined as follows.

1. *$f(X^i p) := p_i \in \Phi_i$ $(i \in \omega)$ for any $p \in \Phi$ (especially, $f(p) := p \in \Phi$),*
2. *$f(X^i(\alpha \circ \beta)) := f(X^i \alpha) \circ f(X^i \beta)$ where $\circ \in \{\to, \wedge, \vee\}$,*
3. *$f(X^i \neg \alpha) := \neg f(X^i \alpha)$,*
4. *$f(X^i G\alpha) := \bigwedge \{f(X^{i+j}\alpha) \mid j \in \omega\}$,*
5. *$f(X^i F\alpha) := \bigvee \{f(X^{i+j}\alpha) \mid j \in \omega\}$.*

An expression $f(\Gamma)$ denotes the result of replacing every occurrence of a formula α in Γ by an occurrence of $f(\alpha)$.

In this definition, the conditions 4 and 5 respectively correspond to the axiom schemes: $G\alpha \leftrightarrow \bigwedge_{i \in \omega} X^i \alpha$ and $F\alpha \leftrightarrow \bigvee_{i \in \omega} X^i \alpha$, which mean "G and F in LTL can respectively be represented by $\bigwedge$ and $\bigvee$ in IL."

2.4   Syntactical Embedding Theorem

Theorem 5 (Syntactical embedding theorem). *Let Γ and Δ be sets of formulas in $\mathcal{L}_{LTL}$, and f be the mapping defined in Definition 4.*

1. *$LT_\omega \vdash \Gamma \Rightarrow \Delta$ iff $LK_\omega \vdash f(\Gamma) \Rightarrow f(\Delta)$.*
2. *$LT_\omega - (cut) \vdash \Gamma \Rightarrow \Delta$ iff $LK_\omega - (cut^0) \vdash f(\Gamma) \Rightarrow f(\Delta)$.*

Proof. Since the case (2) can be obtained as the subproof of the case (1), we show only (1) in the following.

• ($\Longrightarrow$) : By induction on the proof P of $\Gamma \Rightarrow \Delta$ in LT_ω. We distinguish the cases according to the last inference of P, and show some cases.

Case ($X^i p \Rightarrow X^i p$): The last inference of P is of the form: $X^i p \Rightarrow X^i p$. In this case, we obtain $LK_\omega \vdash f(X^i p) \Rightarrow f(X^i p)$, i.e., $LK_\omega \vdash p_i \Rightarrow p_i$ $(p_i \in \Phi_i)$, since $p_i \Rightarrow p_i$ is an initial sequent of LK_ω.

Case ($\to$left): The last inference of P is of the form:

$$\frac{\Gamma \Rightarrow \Sigma, X^i \alpha \quad X^i \beta, \Delta \Rightarrow \Pi}{X^i(\alpha \to \beta), \Gamma, \Delta \Rightarrow \Sigma, \Pi} \; (\to\text{left}).$$

By the hypothesis of induction, we have $LK_\omega \vdash f(\Gamma) \Rightarrow f(\Sigma), f(X^i \alpha)$ and $LK_\omega \vdash f(X^i \beta), f(\Delta) \Rightarrow f(\Pi)$. Then, we obtain

$$\frac{f(\Gamma) \Rightarrow f(\Sigma), f(X^i \alpha) \quad f(X^i \beta), f(\Delta) \Rightarrow f(\Pi)}{f(X^i \alpha) \to f(X^i \beta), f(\Gamma), f(\Delta) \Rightarrow f(\Sigma), f(\Pi)} \; (\to\text{left}^0)$$

where $f(\mathrm{X}^i\alpha){\to}f(\mathrm{X}^i\beta) = f(\mathrm{X}^i(\alpha{\to}\beta))$.

Case (Gleft): The last inference of P is of the form:

$$\frac{\mathrm{X}^{i+k}\alpha, \Gamma \Rightarrow \Delta}{\mathrm{X}^i G\alpha, \Gamma \Rightarrow \Delta} \text{ (Gleft).}$$

By the hypothesis of induction, we have $\mathrm{LK}_\omega \vdash f(\mathrm{X}^{i+k}\alpha), f(\Gamma) \Rightarrow f(\Delta)$, and hence obtain:

$$\vdots$$

$$\frac{f(\mathrm{X}^{i+k}\alpha), f(\Gamma) \Rightarrow f(\Delta) \qquad (f(\mathrm{X}^{i+k}\alpha) \in \{f(\mathrm{X}^{i+j}\alpha) \mid j \in \omega\})}{\bigwedge\{f(\mathrm{X}^{i+j}\alpha) \mid j \in \omega\}, f(\Gamma) \Rightarrow f(\Delta)} \text{ } (\bigwedge \text{left})$$

where $\bigwedge\{f(\mathrm{X}^{i+j}\alpha) \mid j \in \omega\} = f(\mathrm{X}^i G\alpha)$.

Case (Gright): The last inference of P is of the form:

$$\frac{\{\ \Gamma \Rightarrow \Delta, \mathrm{X}^{i+j}\alpha\ \}_{j\in\omega}}{\Gamma \Rightarrow \Delta, \mathrm{X}^i G\alpha} \text{ (Gright).}$$

By the hypothesis of induction, we have $\mathrm{LK}_\omega \vdash f(\Gamma) \Rightarrow f(\Delta), f(\mathrm{X}^{i+j}\alpha)$ for all $j \in \omega$. Let Φ be $\{f(\mathrm{X}^{i+j}\alpha) \mid j \in \omega\}$. We obtain

$$\vdots$$

$$\frac{\{\ f(\Gamma) \Rightarrow f(\Delta), f(\mathrm{X}^{i+j}\alpha)\ \}_{f(\mathrm{X}^{i+j}\alpha)\in\Phi}}{f(\Gamma) \Rightarrow f(\Delta), \bigwedge \Phi} \text{ } (\bigwedge \text{right})$$

where $\bigwedge \Phi = f(\mathrm{X}^i G\alpha)$.

• ($\Longleftarrow$) : By induction on the proof Q of $f(\Gamma) \Rightarrow f(\Delta)$ in LK_ω. We distinguish the cases according to the last inference of Q, and show only the following case.

Case ($\bigwedge$right): The last inference of Q is of the form:

$$\frac{\{\ f(\Gamma) \Rightarrow f(\Delta), f(\mathrm{X}^{i+j}\alpha)\ \}_{f(\mathrm{X}^{i+j}\alpha)\in\Phi}}{f(\Gamma) \Rightarrow f(\Delta), \bigwedge \Phi} \text{ } (\bigwedge \text{right})$$

where $\Phi = \{f(\mathrm{X}^{i+j}\alpha) \mid j \in \omega\}$ and $\bigwedge \Phi = f(\mathrm{X}^i G\alpha)$. By the hypothesis of induction, we have $\mathrm{LT}_\omega \vdash \Gamma \Rightarrow \Delta, \mathrm{X}^{i+j}\alpha$ for all $j \in \omega$. We thus obtain the required fact:

$$\vdots$$

$$\frac{\{\ \Gamma \Rightarrow \Delta, \mathrm{X}^{i+j}\alpha\ \}_{j\in\omega}}{\Gamma \Rightarrow \Delta, \mathrm{X}^i G\alpha} \text{ (Gright).} \qquad \blacksquare$$

3   Semantical Embedding

3.1   Semantics of LTL

For the sake of brevity, the implication connective $\rightarrow$ is omitted in the language of the semantics of LTL. The notations and terminologies are not changed from the previous section. The symbol $\geq$ or $\leq$ is used to represent a linear order on ω. In the following, LTL is defined semantically as a consequence relation. The semantics of LTL is also called LTL. The following definition is from [17].

Definition 6 (LTL). *Let S be a non-empty set of states. A structure (σ, I) is a model if*

1. *σ is an infinite sequence $s_0, s_1, s_2, \ldots$ of states in S,*
2. *I is a mapping from the set Φ of propositional variables to the power set of S.*

In this definition, σ is called a computation, *and I is called an* interpretation.
A consequence relation $(\sigma, I, i) \models_{LTL} \alpha$ for any formula α, where (σ, I) is a model, and i $(\in \omega)$ represents some position within σ, is defined inductively by

1. *$(\sigma, I, i) \models_{LTL} p$ iff $s_i \in I(p)$ for any $p \in \Phi$,*
2. *$(\sigma, I, i) \models_{LTL} \alpha \wedge \beta$ iff $(\sigma, I, i) \models_{LTL} \alpha$ and $(\sigma, I, i) \models_{LTL} \beta$,*
3. *$(\sigma, I, i) \models_{LTL} \alpha \vee \beta$ iff $(\sigma, I, i) \models_{LTL} \alpha$ or $(\sigma, I, i) \models_{LTL} \beta$,*
4. *$(\sigma, I, i) \models_{LTL} \neg\alpha$ iff not-$[(\sigma, I, i) \models_{LTL} \alpha]$,*
5. *$(\sigma, I, i) \models_{LTL} X\alpha$ iff $(\sigma, I, i+1) \models_{LTL} \alpha$,*
6. *$(\sigma, I, i) \models_{LTL} G\alpha$ iff $(\sigma, I, j) \models_{LTL} \alpha$ for any $j \geq i$,*
7. *$(\sigma, I, i) \models_{LTL} F\alpha$ iff $(\sigma, I, j) \models_{LTL} \alpha$ for some $j \geq i$.*

A formula α is valid in LTL if $(\sigma, I, 0) \models_{LTL} \alpha$ for any model (σ, I).

3.2   Semantics of IL

For the sake of brevity, the implication connective $\rightarrow$ is omitted in the language of the semantics of IL. The notations and terminologies are not changed from the previous section. The semantics of IL is also called IL in the following.

Definition 7 (IL). *Let Θ be a countable (non-empty) set of formulas. V is a mapping from the set Φ of propositional variables to the set $\{t, f\}$ of truth values. V is called a* valuation. *A valuation V is extended to a mapping from the set of formulas to $\{t, f\}$ by*

1. *$V(\neg\alpha) = t$ iff $V(\alpha) = f$,*
2. *$V(\bigwedge \Theta) = t$ iff $V(\alpha) = t$ for all $\alpha \in \Theta$,*
3. *$V(\bigvee \Theta) = t$ iff $V(\alpha) = t$ for some $\alpha \in \Theta$.*

In order to make a comparison between LTL and IL, a consequence relation $V \models_{IL} \alpha$ for any formula α is inductively defined by

1. $V \models_{IL} p$ iff $V(p) = t$ for any $p \in \Phi$,
2. $V \models_{IL} \neg\alpha$ iff not-$(V \models_{IL} \alpha)$,
3. $V \models_{IL} \bigwedge \Theta$ iff $V \models_{IL} \alpha$ for any $\alpha \in \Theta$,
4. $V \models_{IL} \bigvee \Theta$ iff $V \models_{IL} \alpha$ for some $\alpha \in \Theta$.

A formula α is valid in IL if $V \models_{IL} \alpha$ (or equivalently $V(\alpha) = t$) for any valuation V.

3.3 Semantical Embedding Theorem

To prove the semantical embedding theorem, the following lemma is shown.

Lemma 8 Let Φ, Φ_i ($i \in \omega$) and f be the same as those in Definition 4 (but $\rightarrow$ is deleted in the definition). Suppose that V is a valuation from $\bigcup_{i \in \omega} \Phi_i$ to $\{t, f\}$, S is a non-empty set of states, and (σ, I) is a model such that σ is a computation $s_0, s_1, s_2, \dots$ ($s_i \in S, i \in \omega$), and I is an interpretation from Φ to the power set of S satisfying the following condition:

$$\forall i \in \omega, \forall p \in \Phi \ [s_i \in I(p) \ \text{iff} \ V(p_i) = t] .$$

Then, for any formula α in $\mathcal{L}_{LTL}$,

$$(\sigma, I, i) \models_{LTL} \alpha \ \text{iff} \ V \models_{IL} f(X^i \alpha).$$

Proof. By induction on the complexity of α. For the sake of simplicity, V of $V \models_{IL}$ is omitted in the following.

- Base step: $\alpha \equiv p \in \Phi$. $(\sigma, I, i) \models_{LTL} p$ iff $s_i \in I(p)$ iff $V(p_i) = t$ iff $\models_{IL} p_i$ iff $\models_{IL} f(X^i p)$.
- Induction step.

(Case $\alpha \equiv \beta \wedge \gamma$): $(\sigma, I, i) \models_{LTL} \beta \wedge \gamma$ iff $(\sigma, I, i) \models_{LTL} \beta$ and $(\sigma, I, i) \models_{LTL} \gamma$ iff $\models_{IL} f(X^i \beta)$ and $\models_{IL} f(X^i \gamma)$ (by the induction hypothesis) iff $\models_{IL} f(X^i \beta) \wedge f(X^i \gamma)$ iff $\models_{IL} f(X^i(\beta \wedge \gamma))$.

(Case $\alpha \equiv \beta \vee \gamma$): Similar to the above case.

(Case $\alpha \equiv \neg\beta$): $(\sigma, I, i) \models_{LTL} \neg\beta$ iff not-$[(\sigma, I, i) \models_{LTL} \beta]$ iff not-$[\models_{IL} f(X^i \beta)]$ (by the induction hypothesis) iff $\models_{IL} \neg f(X^i \beta)$ iff $\models_{IL} f(X^i \neg\beta)$.

(Case $\alpha \equiv X\beta$): $(\sigma, I, i) \models_{LTL} X\beta$ iff $(\sigma, I, i+1) \models_{LTL} \beta$ iff $\models_{IL} f(X^{i+1} \beta)$ (by the induction hypothesis) iff $\models_{IL} f(X^i(X\beta))$.

(Case $\alpha \equiv G\beta$): $(\sigma, I, i) \models_{LTL} G\beta$ iff $\forall j \geq i \ [(\sigma, I, j) \models_{LTL} \beta]$ iff $\forall j \geq i \ [\models_{IL} f(X^j \beta)]$ (by the induction hypothesis) iff $\forall k \in \omega \ [\models_{IL} f(X^{i+k} \beta)]$ iff $\models_{IL} \gamma$ for all $\gamma \in \{f(X^{i+k} \beta) \mid k \in \omega\}$ iff $\models_{IL} \bigwedge\{f(X^{i+k} \beta) \mid k \in \omega\}$ iff $\models_{IL} f(X^i G\beta)$.

(Case $\alpha \equiv F\beta$): Similar to the above case. ∎

We then obtain the following theorem as a special case.

Theorem 9 (Semantical embedding theorem). Let f be the mapping defined in Definition 4. Then, for any formula α in $\mathcal{L}_{LTL}$,

$$\alpha \ \text{is valid in LTL iff} \ f(\alpha) \ \text{is valid in IL.}$$

4 Applications to Multi-agent Sequent Systems

4.1 Cut-Elimination for LT$_\omega$ and ILT$_\omega$

The cut-elimination theorem for LT$_\omega$ is easily obtained as an application of Theorem 5.

Theorem 10 (Cut-elimination for LT$_\omega$). *The rule* (cut) *is admissible in cut-free* LT$_\omega$.

Proof. Suppose LT$_\omega \vdash \Gamma \Rightarrow \Delta$. Then, we have LK$_\omega \vdash f(\Gamma) \Rightarrow f(\Delta)$ by Theorem 5 (1), and hence LK$_\omega - (\text{cut}^0) \vdash f(\Gamma) \Rightarrow f(\Delta)$ by Theorem 3. By Theorem 5 (2), we obtain LT$_\omega - (\text{cut}) \vdash \Gamma \Rightarrow \Delta$. ∎

An *infinitary linear-time temporal logic* (ILT$_\omega$), which is an integration of LT$_\omega$ and LK$_\omega$, is introduced below.

Definition 11 (ILT$_\omega$). ILT$_\omega$ *is obtained from* LT$_\omega$ *by deleting* $\{(\wedge\text{left1}),$ $(\wedge\text{left2}), (\wedge\text{right}), (\vee\text{left}), (\vee\text{right1}), (\vee\text{right2})\}$ *and adding the inference rules of the form:*

$$\frac{X^i\alpha, \Gamma \Rightarrow \Delta \ (\alpha \in \Theta)}{X^i(\bigwedge\Theta), \Gamma \Rightarrow \Delta} \ (\textstyle\bigwedge\text{left}*) \qquad \frac{\{\ \Gamma \Rightarrow \Delta, X^i\alpha\ \}_{\alpha\in\Theta}}{\Gamma \Rightarrow \Delta, X^i(\bigwedge\Theta)} \ (\textstyle\bigwedge\text{right}*)$$

$$\frac{\{\ X^i\alpha, \Gamma \Rightarrow \Delta\ \}_{\alpha\in\Theta}}{X^i(\bigvee\Theta), \Gamma \Rightarrow \Delta} \ (\textstyle\bigvee\text{left}*) \qquad \frac{\Gamma \Rightarrow \Delta, X^i\alpha \ (\alpha \in \Theta)}{\Gamma \Rightarrow \Delta, X^i(\bigvee\Theta)} \ (\textstyle\bigvee\text{right}*)$$

where Θ *denotes a non-empty countable set of formulas.*

Proposition 12. *The following sequents are provable in cut-free* ILT$_\omega$:

1. $G\alpha \Leftrightarrow \bigwedge\{X^i\alpha \mid i \in \omega\}$,
2. $F\alpha \Leftrightarrow \bigvee\{X^i\alpha \mid i \in \omega\}$.

By using an appropriate modification of Theorem 5, the following theorem can simply be obtained. This theorem is a new result of this paper. The direct Gentzen-style proof of this theorem may be very difficult and complex. Compared with the direct proof, our embedding-based method is very easy and simple. This is a merit of our method.

Theorem 13 (Cut-elimination for ILT$_\omega$). *The rule* (cut) *is admissible in cut-free* ILT$_\omega$.

Theorems 10 and 13 can also be extended to the theorems for the first-order versions. These extended theorems can be obtained from Theorem 20 discussed in the next subsection.

4.2 Cut-Elimination for IELT$_\omega$

In this subsection, a *first-oeder multi-agent infinitary epistemic linear-time temporal logic* (IELT$_\omega$) is introduced as a sequent calculus by integrating and extending both ILT$_\omega$ and a *multi-agent infinitary epistemic logic* S4$_\omega$. The following list of symbols is adopted for the language of the underlying logic: free variables $a_0, a_1, ...$, bound variables $x_0, x_1, ...$, functions $f_0, f_1, ...$, predicates $p_0, p_1, ...$, logical connectives $\rightarrow, \neg, \wedge, \vee, \forall$ (any), $\exists$ (exists), knowledge operators $K_1, K_2, ..., K_n$ (agent i knows), temporal operators G, F, X. The numbers of free and bound variables are assumed to be countable, and the numbers of functions and predicates are also assumed to be countable. It is also assumed there is at least one predicate. A 0-ary function is an individual constant, and a 0-ary predicate is a propositional variable. The total number n of the knowledge operators is assumed to be an arbitrary fixed positive integer. Then, the following definition is from [12].

Definition 14. *Assume that the notion of* term *is defined by the standard way. Let F_0 be the set of all formulas generated by the standard finitely inductive definition with respect to $\{\rightarrow, \neg, \forall, \exists, K_1, ..., K_n, X, G, F\}$ from the set of atomic formulas. Suppose that F_t is already defined with respect to $t = 0, 1, 2, ...$ A non-empty countable subset Θ of F_t is called an* allowable set *if it contains a finite number of free variables. The expressions $\bigwedge \Theta$ and $\bigvee \Theta$ for an allowable set Θ are considered below. Let Θ_t be an allowable set in F_t. We define F_{t+1} from $F_t \cup \{\bigwedge \Theta_t, \bigwedge \Theta_t\}$ by the standard finitely inductive definition with respect to $\{\rightarrow, \neg, \forall, \exists, K_1, ..., K_n, X, G, F\}$. F_ω, which is called the set of formulas, is defined by $\bigcup_{t < \omega} F_t$, and an expression in F_ω is called a* formula. *An expression of the form $\Gamma \Rightarrow \Delta$ where Γ and Δ are finite (possibly empty) sets of formulas is called a* sequent.

The notations used in this subsection are almost the same as those of the previous ones. For any $\sharp \in \{K_1, K_2, ..., K_n\}$, an expression $\sharp\Gamma$ is used to denote the set $\{\sharp\gamma \mid \gamma \in \Gamma\}$.

Definition 15 (IELT$_\omega$). *IELT$_\omega$ is obtained from ILT$_\omega$ by deleting $\{(\bigwedge \mathrm{left}*), (\bigwedge \mathrm{right}*), (\bigvee \mathrm{left}*), (\bigvee \mathrm{right}*)\}$ and adding the inference rules of the form:*

$$\frac{X^i\alpha, \Gamma \Rightarrow \Delta \quad (\alpha \in \Theta)}{X^i(\bigwedge \Theta), \Gamma \Rightarrow \Delta} \ (\bigwedge \mathrm{left}\star) \qquad \frac{\{\ \Gamma \Rightarrow \Delta, X^i\alpha\ \}_{\alpha \in \Theta}}{\Gamma \Rightarrow \Delta, X^i(\bigwedge \Theta)} \ (\bigwedge \mathrm{right}\star)$$

$$\frac{\{\ X^i\alpha, \Gamma \Rightarrow \Delta\ \}_{\alpha \in \Theta}}{X^i(\bigvee \Theta), \Gamma \Rightarrow \Delta} \ (\bigvee \mathrm{left}\star) \qquad \frac{\Gamma \Rightarrow \Delta, X^i\alpha \quad (\alpha \in \Theta)}{\Gamma \Rightarrow \Delta, X^i(\bigvee \Theta)} \ (\bigvee \mathrm{right}\star)$$

where Θ is an allowable set,

$$\frac{X^i\alpha(t), \Gamma \Rightarrow \Delta}{X^i\forall x\alpha(x), \Gamma \Rightarrow \Delta} \ (\forall \mathrm{left}) \qquad \frac{\Gamma \Rightarrow \Delta, X^i\alpha(a)}{\Gamma \Rightarrow \Delta, X^i\forall x\alpha(x)} \ (\forall \mathrm{right})$$

$$\frac{X^i\alpha(a), \Gamma \Rightarrow \Delta}{X^i\exists x\alpha(x), \Gamma \Rightarrow \Delta} \ (\exists\text{left}) \qquad \frac{\Gamma \Rightarrow \Delta, X^i\alpha(t)}{\Gamma \Rightarrow \Delta, X^i\exists x\alpha(x)} \ (\exists\text{right})$$

where a is a free variable which must not occur in the lower sequents of ($\forall$right)
and ($\exists$left), *and t is an arbitrary term,*

$$\frac{X^i\alpha, \Gamma \Rightarrow \Delta}{X^iK_m\alpha, \Gamma \Rightarrow \Delta} \ (K_m\text{left}) \qquad \frac{X^iK_m\Gamma \Rightarrow X^k\alpha}{X^iK_m\Gamma \Rightarrow X^kK_m\alpha} \ (K_m\text{right})$$

where $m \in \{1, ..., n\}$.

A proof *in* IELT$_\omega$ *is defined as a tree similarly in* [12]. *A sequent S is called provable in* IELT$_\omega$ *if there is a proof of S in* IELT$_\omega$. *The same notions for proofs and provabilities are assumed to be defined similarly for some subsystems of* IELT$_\omega$.

Definition 16 (S4$_\omega$). *A sequent calculus* S4$_\omega$ *for an infinitary (and multimodal) version of the modal logic* S4 *is obtained from* IELT$_\omega$ *by deleting* {(Gleft), (Gright), (Fleft), (Fright)} *and replacing* i, k *by* 0 *(i.e., deleting every occurrence of* X*). The modified inference rules for* S4$_\omega$ *by replacing* i, k *by* 0 *are denoted by labeling "*0*" in superscript, e.g.,* ($\rightarrow$left^0) *and* ($\bigwedge$left$\star^0$).

The following cut-elimination theorem is known (see e.g., [29,12,28]).

Theorem 17. *The rule* (cut^0) *is admissible in cut-free* S4$_\omega$.

Definition 18. *We fix a countable non-empty set* Φ *of atomic formulas, and define the sets* $\Phi_i := \{p_i \mid p \in \Phi\}$ ($1 \le i \in \omega$) *and* $\Phi_0 := \Phi$ *of atomic formulas. The language* $\mathcal{L}_{\text{IELT}_\omega}$ *(or the set of formulas) of* IELT$_\omega$ *is defined by using* Φ, $\rightarrow, \neg, \bigwedge, \bigvee, \forall, \exists,$ K$_1, ...,$ K$_n$, X, G *and* F *in the same way as that in Definition 14. The language* $\mathcal{L}_{\text{S4}_\omega}$ *of* S4$_\omega$ *is defined by using* $\bigcup_{i\in\omega}\Phi_i$, $\rightarrow, \neg,$ $\bigwedge, \bigvee, \forall, \exists,$ K$_1, ...,$ K$_n$ *in a similar way as that in Definition 14.*

A mapping f *from* $\mathcal{L}_{\text{IELT}_\omega}$ *to* $\mathcal{L}_{\text{S4}_\omega}$ *is defined by the same conditions 1–5 in Definition 4 and adding the following conditions:*

1. $f(X^i(\sharp\Theta)) := \sharp f(X^i\Theta)$ *where* Θ *is an allowable set,* $\sharp \in \{\bigwedge, \bigvee\}$, *and* $f(X^i\Theta)$ *is the result of replacing every occurrence of a formula* α *in* $X^i\Theta$ *by an occurrence of* $f(\alpha)$,
2. $f(X^iQx\alpha(x)) := Qxf(X^i\alpha(x))$ *where* $Q \in \{\forall, \exists\}$,
3. $f(X^iK_m\alpha) := K_mf(X^i\alpha)$ *for any* $m \in \{1, ..., n\}$.

Theorem 19 (Extended syntactical embedding theorem). *Let* Γ *and* Δ *be sets of formulas in* $\mathcal{L}_{\text{IELT}_\omega}$, *and* f *be the mapping defined in Definition 18.*

1. IELT$_\omega \vdash \Gamma \Rightarrow \Delta$ *iff* S4$_\omega \vdash f(\Gamma) \Rightarrow f(\Delta)$.
2. IELT$_\omega$ − (cut) $\vdash \Gamma \Rightarrow \Delta$ *iff* S4$_\omega$ − (cut^0) $\vdash f(\Gamma) \Rightarrow f(\Delta)$.

Proof. We show only the direction ($\Longrightarrow$) of (1) by induction on the proof P of $\Gamma \Rightarrow \Delta$ in IELT$_\omega$. We distinguish the cases according to the last inference of P, and show some cases.

Case ($\bigwedge$right$\star$): The last inference of P is of the form:

$$\frac{\{\ \Gamma \Rightarrow \Delta, \mathrm{X}^i\alpha\ \}_{\alpha \in \Theta}}{\Gamma \Rightarrow \Delta, \mathrm{X}^i(\bigwedge\Theta)}\ (\bigwedge\text{right}\star).$$

By the hypothesis of induction, we have $\mathrm{S4}_\omega \vdash f(\Gamma) \Rightarrow f(\Delta), f(\mathrm{X}^i\alpha)$ for all $\alpha \in \Theta$, i.e., for all $f(\mathrm{X}^i\alpha) \in f(\mathrm{X}^i\Theta)$. Then, we obtain

$$\vdots$$

$$\frac{\{\ f(\Gamma) \Rightarrow f(\Delta), f(\mathrm{X}^i\alpha)\ \}_{f(\mathrm{X}^i\alpha) \in f(\mathrm{X}^i\Theta)}}{f(\Gamma) \Rightarrow f(\Delta), \bigwedge f(\mathrm{X}^i\Theta)}\ (\bigwedge\text{right}\star^0).$$

where $\bigwedge f(\mathrm{X}^i\Theta) = f(\mathrm{X}^i(\bigwedge\Theta))$.

Case ($\mathrm{K}_m$right): The last inference of P is of the form:

$$\frac{\mathrm{X}^i\mathrm{K}_m\Gamma \Rightarrow \mathrm{X}^k\alpha}{\mathrm{X}^i\mathrm{K}_m\Gamma \Rightarrow \mathrm{X}^k\mathrm{K}_m\alpha}\ (\mathrm{K}_m\text{right}).$$

Bt the hypothesis of induction, we have $\mathrm{S4}_\omega \vdash f(\mathrm{X}^i\mathrm{K}_m\Gamma) \Rightarrow f(\mathrm{X}^k\alpha)$, i.e., $\mathrm{S4}_\omega \vdash \mathrm{K}_m f(\mathrm{X}^i\Gamma) \Rightarrow f(\mathrm{X}^k\alpha)$. Then, we obtain

$$\vdots$$

$$\frac{\mathrm{K}_m f(\mathrm{X}^i\Gamma) \Rightarrow f(\mathrm{X}^k\alpha)}{\mathrm{K}_m f(\mathrm{X}^i\Gamma) \Rightarrow \mathrm{K}_m f(\mathrm{X}^k\alpha)}\ (\mathrm{K}_m\text{right}^0).$$

Since $\mathrm{K}_m f(\mathrm{X}^i\Gamma) = f(\mathrm{X}^i\mathrm{K}_m\alpha)$ and $\mathrm{K}_m f(\mathrm{X}^k\alpha) = f(\mathrm{X}^k\mathrm{K}_m\alpha)$, we have the required fact: $\mathrm{S4}_\omega \vdash f(\mathrm{X}^i\mathrm{K}_m\Gamma) \Rightarrow f(\mathrm{X}^k\mathrm{K}_m\alpha)$. $\blacksquare$

Using Theorems 19 and 17, we obtain the following new theorem, which is the main result of this paper. This theorem may not be obtained by any other methods because of the complexity of the proofs. By the virtue of our new embedding-based proof method, we can avoid such complexity.

Theorem 20 (Cut-elimination for IELT$_\omega$). *The rule* (cut) *is admissible in cut-free* IELT$_\omega$.

This theorem guarantees the consistency of IELT$_\omega$ (i.e., the empty sequent $\Rightarrow$ is not provable in IELT$_\omega$) and the conservativity of IELT$_\omega$ (i.e., IELT$_\omega$ is a conservative extension of LT$_\omega$, LK$_\omega$, S4$_\omega$ and ILT$_\omega$).

4.3   Cut-Elimination and Decidability for ELT$_l$

In this subsection, a *propositional multi-agent epistemic bounded linear-time temporal logic* (ELT$_l$), which is regarded as a modified subsystem of IELT$_\omega$, is introduced as a sequent calculus by restricting the time domain by an initial segment ω_l of ω. An approach which is based on a bounded time domain was studied in [10]. In [10], propositional and first-order *bounded linear-time temporal logics* (BLTL and FBLTL, respectively), were introduced by restricting the propositional and first-order versions of LT_ω, and the completeness theorems for BLTL and FBLTL were proved. The embedding theorems of BLTL and FBLTL into (propositional and first-order, respectively) LK for classical logic were also shown in [10]. In the following, a similar embedding result is shown for ELT_l.

The notations used in this subsection are almost the same as those of the previous ones.

Definition 21 (ELT_l). *Let l be a fixed positive integer, and ω_l be $\{i \in \omega \mid i \le l\}$.*

ELT_l *is obtained from the propositional fragment of IELT_ω by deleting* $\{(\bigwedge\mathrm{left}\star), (\bigwedge\mathrm{right}\star), (\bigvee\mathrm{left}\star), (\bigvee\mathrm{right}\star), (\mathrm{Gleft}), (\mathrm{Gright}), (\mathrm{Fleft}), (\mathrm{Fright})\}$ *and adding* $\{(\wedge\mathrm{left}1), (\wedge\mathrm{left}2), (\wedge\mathrm{right}), (\vee\mathrm{left}), (\vee\mathrm{right}1), (\vee\mathrm{right}2)\}$ *and the inference rules of the form: for any $k \in \omega_l$,*

$$\frac{\mathrm{X}^l\alpha, \Gamma \Rightarrow \Delta}{\mathrm{X}^{i+l}\alpha, \Gamma \Rightarrow \Delta} \ (\mathrm{Xleft}) \qquad \frac{\Gamma \Rightarrow \Delta, \mathrm{X}^l\alpha}{\Gamma \Rightarrow \Delta, \mathrm{X}^{i+l}\alpha} \ (\mathrm{Xright})$$

$$\frac{\mathrm{X}^{i+k}\alpha, \Gamma \Rightarrow \Delta}{\mathrm{X}^i\mathrm{G}\alpha, \Gamma \Rightarrow \Delta} \ (\mathrm{Gleft}^l) \qquad \frac{\{\, \Gamma \Rightarrow \Delta, \mathrm{X}^{i+j}\alpha \,\}_{j\in\omega_l}}{\Gamma \Rightarrow \Delta, \mathrm{X}^i\mathrm{G}\alpha} \ (\mathrm{Gright}^l)$$

$$\frac{\{\, \mathrm{X}^{i+j}\alpha, \Gamma \Rightarrow \Delta \,\}_{j\in\omega_l}}{\mathrm{X}^i\mathrm{F}\alpha, \Gamma \Rightarrow \Delta} \ (\mathrm{Fleft}^l) \qquad \frac{\Gamma \Rightarrow \Delta, \mathrm{X}^{i+k}\alpha}{\Gamma \Rightarrow \Delta, \mathrm{X}^i\mathrm{F}\alpha} \ (\mathrm{Fright}^l).$$

It is remarked that ELT_l is regarded as a multi-agent epistemic extension of the bounded time version of LT_ω.

Definition 22 ($\mathrm{S4}_n$). *A sequent calculus $\mathrm{S4}_n$ for a multi-modal version of the modal logic S4 is obtained from ELT_l by deleting* $\{(\mathrm{Xleft}), (\mathrm{Xright}), (\mathrm{Gleft}^l), (\mathrm{Gright}^l), (\mathrm{Fleft}^l), (\mathrm{Fright}^l)\}$ *and replacing i, k by 0. The modified inference rules for $\mathrm{S4}_n$ by replacing i, k by 0 are denoted by labeling "0" in superscript.*

It is noted that $\mathrm{S4}_n$ is regarded as a finite and propositional subsystem of $\mathrm{S4}_\omega$.

The following is known.

Theorem 23. *The following facts hold:*

1. *The rule (cut^0) is admissible in cut-free $\mathrm{S4}_n$.*
2. *$\mathrm{S4}_n$ is decidable.*

In the following, expressions like $\bigwedge\{\alpha_i \mid i \in \omega_l\}$ and $\bigvee\{\alpha_i \mid i \in \omega_l\}$ where $\{\alpha_i \mid i \in \omega_l\}$ is a multiset mean $\alpha_0 \wedge \alpha_1 \wedge \cdots \wedge \alpha_l$ and $\alpha_0 \vee \alpha_1 \vee \cdots \vee \alpha_l$, respectively. For example, $\bigvee\{\alpha, \alpha, \beta\}$ means $\alpha \vee \alpha \vee \beta$.

Definition 24. *We fix a countable non-empty set Φ of atomic formulas, and define the sets $\Phi_i := \{p_i \mid p \in \Phi\}$ ($1 \le i \le \omega$) and $\Phi_0 := \Phi$ of atomic formulas. The language $\mathcal{L}_{\mathrm{ELT}_l}$ of ELT_l is defined by using $\Phi, \to, \neg, \wedge, \vee, \mathrm{K}_1, ..., \mathrm{K}_n, \mathrm{X}, \mathrm{G}$ and F. The language $\mathcal{L}_{\mathrm{S4}_n}$ of $\mathrm{S4}_n$ is defined by using $\bigcup_{i\in\omega} \Phi_i, \to, \neg, \wedge, \vee$ and $\mathrm{K}_1, ..., \mathrm{K}_n$.*

A mapping f from $\mathcal{L}_{\mathrm{ELT}_l}$ to $\mathcal{L}_{\mathrm{S4}_n}$ is obtained from the conditions of Definition 4 by deleting the conditions 4 and 5, and adding the following conditions:

1. $f(X^i K_m \alpha) := K_m f(X^i \alpha)$ *for any* $m \in \{1, ..., n\}$,
2. $f(X^{i+l} \alpha) := f(X^l \alpha)$,
3. $f(X^i G \alpha) := \bigwedge \{ f(X^{i+j} \alpha) \mid j \in \omega_l \}$,
4. $f(X^i F \alpha) := \bigvee \{ f(X^{i+j} \alpha) \mid j \in \omega_l \}$.

Theorem 25 (Modified syntactical embedding theorem). *Let Γ and Δ be sets of formulas in $\mathcal{L}_{\mathrm{ELT}_1}$, and f be the mapping defined in Definition 24.*

1. $\mathrm{ELT}_l \vdash \Gamma \Rightarrow \Delta$ *iff* $\mathrm{S4}_n \vdash f(\Gamma) \Rightarrow f(\Delta)$.
2. $\mathrm{ELT}_l - (\mathrm{cut}) \vdash \Gamma \Rightarrow \Delta$ *iff* $\mathrm{S4}_n - (\mathrm{cut}^0) \vdash f(\Gamma) \Rightarrow f(\Delta)$.

Proof. We show only the direction ($\Longrightarrow$) of (1) by induction on a proof P of $\Gamma \Rightarrow \Delta$ in ELT_l. We distinguish the cases according to the last inference of P, and show some cases.

Case (Xleft): The last inference of P is of the form:

$$\frac{X^l \alpha, \Gamma \Rightarrow \Delta}{X^{i+l} \alpha, \Gamma \Rightarrow \Delta} \ (\text{Xleft}).$$

By the hypothesis of induction, we have $\mathrm{LK} \vdash f(X^l \alpha), f(\Gamma) \Rightarrow f(\Delta)$, and $f(X^l \alpha) = f(X^{i+l} \alpha)$. Thus, we obtain $\mathrm{LK} \vdash f(X^{i+l} \alpha), f(\Gamma) \Rightarrow f(\Delta)$.

Case (Gleft^l): The last inference of P is of the form:

$$\frac{X^{i+k} \alpha, \Gamma \Rightarrow \Delta}{X^i G \alpha, \Gamma \Rightarrow \Delta} \ (\text{Gleft}^l).$$

By the hypothesis of induction, we have $\mathrm{LK} \vdash f(X^{i+k} \alpha), f(\Gamma) \Rightarrow f(\Delta)$, and hence obtain:

$$\begin{array}{c} \vdots \\ f(X^{i+k} \alpha), f(\Gamma) \Rightarrow f(\Delta) \\ \vdots \ (\wedge\text{left}^0) \\ \bigwedge \{ f(X^{i+j} \alpha) \mid j \in \omega_l \}, f(\Gamma) \Rightarrow f(\Delta) \end{array}$$

where $\bigwedge \{ f(X^{i+j} \alpha) \mid j \in \omega_l \} = f(X^i G \alpha)$, and $f(X^{i+k} \alpha) \in \{ f(X^{i+j} \alpha) \mid j \in \omega_l \}$. In this proof, the case $i > l$ is also included. In such a case, $f(X^{i+k} \alpha)$ and

$$\bigwedge \{ f(X^{i+j} \alpha) \mid j \in \omega_l \} \text{ mean } f(X^l \alpha) \text{ and } \overbrace{f(X^l \alpha) \wedge f(X^l \alpha) \wedge \cdots \wedge f(X^l \alpha)}^{l}. \ \blacksquare$$

By using Theorems 25 and 23, we obtain the following theorems.

Theorem 26 (Cut-elimination for ELT_l). *The rule* (cut) *is admissible in cut-free ELT_l.*

Theorem 27 (Decidability for ELT_l). ELT_l *is decidable.*

Proof. By Theorem 25, the provability of ELT_l can be transformed into that of $\mathrm{S4}_n$. Since $\mathrm{S4}_n$ is decidable by Theorem 23, ELT_l is also decidable. $\blacksquare$

It is remarked that the same cut-elimination and decidability results can be obtained for the bounded time version of LT_ω.

5 Concluding Remarks

In this paper, the syntactical and semantical embedding theorems of LTL into IL were presented. As applications of the syntactical embedding theorem and its extensions and modifications, the cut-elimination theorems for LT_ω, ILT_ω (infinitary LTL), $IELT_\omega$ (first-order multi-agent infinitary epistemic LTL) and ELT_l (multi-agent epistemic bounded LTL) were shown as the main results of this paper. The embedding theorems give the following slogan: "G and F in LTL are almost (countable) $\bigwedge$ and $\bigvee$ in IL, respectively", in other words, "LTL is almost (the countable fragment of) IL." We hope that this perspective will be useful for developing and combining the research fields of both LTL and IL. In this respect, it was also shown in this paper that our simple and easy embedding method is applicable to the cut-elimination theorems for a wide range of very expressive multi-agent logics combined with LT_ω and LK_ω.

In the following, as an example of the applicability of our method, it is finally remarked that we can obtain the cut-elimination theorems for the modified K4-type, KT-type and K-type versions of $IELT_\omega$ in a similar way. Consider the inference rules of the form:

$$\frac{X^i\Gamma \Rightarrow X^k\alpha}{X^iK_m\Gamma \Rightarrow X^kK_m\alpha} \ (K_mK) \qquad \frac{X^iK_m\Gamma, X^i\Gamma \Rightarrow X^k\alpha}{X^iK_m\Gamma \Rightarrow X^kK_m\alpha} \ (K_mK4).$$

The K4-type version is obtained from $IELT_\omega$ by replacing $\{(K_m\text{left}), (K_m\text{right})\}$ by (K_mK4), the KT-type version is obtained from $IELT_\omega$ by replacing $(K_m\text{left})$ by (K_mK), and the K-type version is obtained from $IELT_\omega$ by replacing $\{(K_m\text{left}), (K_m\text{right})\}$ by (K_mK).

Acknowledgments. This research was supported by the Alexander von Humboldt Foundation. I am grateful to the Foundation for providing excellent working conditions and generous support of this research. This work was partially supported by the Japanese Ministry of Education, Culture, Sports, Science and Technology, Grant-in-Aid for Young Scientists (B) 20700015, 2008.

References

1. Baratella, S., Masini, A.: An approach to infinitary temporal proof theory. Archive for Mathematical Logic 43(8), 965–990 (2004)
2. van Benthem, J.: Modality, bisimulation and interpolation in infinitary logic. Annals of Pure and Applied Logic 96(1-3), 29–41 (1999)
3. Dixon, C., Gago, M.-C.F., Fisher, M., van der Hoek, W.: Using temporal logics of knowledge in the formal verification of security protocols. In: Proceedings of the 11th International Symposium on Temporal Representation and Reasoning (TIME 2004), pp. 148–151. IEEE Computer Society Press, Los Alamitos (2004)
4. Clarke, E.M., Grumberg, O., Peled, D.A.: Model checking. MIT Press, Cambridge (1999)

5. Emerson, E.A.: Temporal and modal logic. In: van Leeuwen, J. (ed.) Handbook of Theoretical Computer Science, Formal Models and Semantics (B), pp. 995–1072. Elsevier and MIT Press (1990)
6. Feferman, S.: Lectures on proof theory. In: Proceedings of the summer school in logic. Lecture Notes in Mathematics, vol. 70, pp. 1–107. Springer, Heidelberg (1968)
7. Halpern, J.H., Moses, Y.: A guide to completeness and complexity for modal logics of knowledge and beliefs. Artificial Intelligence 54, 319–379 (1992)
8. van der Hoek, W., Wooldridge, M.: Cooperation, knowledge and time: Alternating-time temporal epistemic logic and its applications. Studia Logica 75, 125–157 (2003)
9. Kamide, N.: An equivalence between sequent calculi for linear-time temporal logic. Bulletin of the Section of the Logic 35(4), 187–194 (2006)
10. Kamide, N.: Bounded linear-time temporal logic: From Gentzen to Robinson 37 pages (manuscript 2008)
11. Kaneko, M.: Common knowledge logic and game logic. Journal of Symbolic Logic 64(2), 685–700 (1999)
12. Kaneko, M., Nagashima, T.: Game logic and its applications II. Studia Logica 58(2), 273–303 (1997)
13. Kawai, H.: Sequential calculus for a first order infinitary temporal logic. Zeitschrift für Mathematische Logik und Grundlagen der Mathematik 33, 423–432 (1987)
14. Keisler, H.J., Knight, J.F.: Barwise: infinitary logic and addmissible sets. Bulletin of Symbolic Logic 10(1), 4–36 (2004)
15. Kröger, F.: LAR: a logic of algorithmic reasoning. Acta Informatica 8, 243–266 (1977)
16. Lewis, D.K.: Covention: A philosophical study. Harvard University Press (1969)
17. Lichtenstein, O., Pnueli, A.: Propositional temporal logics: decidability and completeness. Logic Journal of the IGPL 8(1), 55–85 (2000)
18. Lismont, L., Mongin, P.: On the logic of common belief and common knowledge. Theory and Decision 37, 75–106 (1994)
19. Lorenzen, P.: Algebraische und logitishe Untersuchungen über freie Verbände. Journal of Symbolic Logic 16, 81–106 (1951)
20. Maruyama, A., Tojo, S., Ono, H.: Temporal epistemic logics for multi-agent models and their efficient proof-search procedures. Computer Software 20(1), 51–65 (2003) (in Japanese)
21. Novikov, P.S.: Inconsistencies of certain logical calculi. In: Infinistic Methods, pp. 71–74. Pergamon, Oxford (1961)
22. Paech, B.: Gentzen-systems for propositional temporal logics. In: Börger, E., Kleine Büning, H., Richter, M.M. (eds.) CSL 1988. LNCS, vol. 385, pp. 240–253. Springer, Heidelberg (1989)
23. Pliuškevičius, R.: Investigation of finitary calculus for a discrete linear time logic by means of infinitary calculus. In: Barzdins, J., Bjorner, D. (eds.) Baltic Computer Science. LNCS, vol. 502, pp. 504–528. Springer, Heidelberg (1991)
24. Pnueli, A.: The temporal logic of programs. In: Proceedings of the 18th IEEE Symposium on Foundations of Computer Science, pp. 46–57 (1977)
25. Szabo, M.E.: A sequent calculus for Kröger logic. In: Salwicki, A. (ed.) Logic of Programs 1980. LNCS, vol. 148, pp. 295–303. Springer, Heidelberg (1983)
26. Szałas, A.: Concerning the semantic consequence relation in first-order temporal logic. Theoretical Computer Science 47(3), 329–334 (1986)
27. Takeuti, G.: Proof theory. North-Holland Pub. Co., Amsterdam (1975)
28. Tanaka, Y.: Cut-elimination theorems for some infinitary modal logics. Mathematical Logic Quarterly 47(3), 327–339 (2001)

29. Tanaka, Y.: Representations of algebras and Kripke completeness of infinitary and predicate logics, Doctor thesis, Japan Advanced Institute of Science and Technology (1999)
30. Wooldridge, M., Dixon, C., Fisher, M.: A tableau-based proof method for temporal logics of knowledge and belief. Journal of Applied Non-Classical Logics 8, 225–258 (1998)
31. Y. Vardi, M.: Branching vs. linear time: final showdown. In: Margaria, T., Yi, W. (eds.) TACAS 2001. LNCS, vol. 2031, pp. 1–22. Springer, Heidelberg (2001)

Bounded-Resource Reasoning
as (Strong or Classical) Planning

Alexandre Albore[1], Natasha Alechina[2], Piergiorgio Bertoli[3],
Chiara Ghidini[3], and Brian Logan[2]

[1] Universitat Pompeu Fabra, pg.Circumval·lació 8 08003, Barcelona, Spain
[2] School of Computer Science, University of Nottingham, Nottingham, NG81BB, UK
[3] FBK-irst, via Sommarive 18, Povo, 38100, Trento, Italy
alexandre.albore@upf.edu,
{nza,bsl}@cs.nott.ac.uk, {bertoli,ghidini}@fbk.eu

Abstract. To appropriately configure agents so as to avoid resource exhaustion, it is necessary to determine the minimum resource (time & memory) requirements necessary to solve reasoning problems. In this paper we show how the problem of reasoning under bounded resources can be recast as a planning problem. Focusing on propositional reasoning, we propose different recasting styles, which are equally interesting, since they require solving different classes of planning problems, and allow representing different reasoner architectures. We implement our approach by automatically encoding problems for the MBP planner. Our experimental results demonstrate that even simple problems can give rise to non-trivial (and often counter intuitive) time and memory saving strategies.

1 Introduction

On devices with limited computational power, the reasoning process of an agent may fail because of lack of memory, or simply take too long to complete. Therefore, to appropriately configure agents and devices which rely on automated reasoning techniques so as to avoid resource exhaustion, we must address the problem of how much time and memory a reasoner embedded in an agent needs to achieve its goals. In this paper we show how the problem of determining the minimal time and memory bounds for propositional reasoning problems can be solved by applying planning techniques.

The key idea of our approach is to model a reasoning agent as a planning domain where fluents correspond to the set of formulas held in the agent's memory, and actions correspond to applications of the agent's inference rules. Naturally, the computational resources (time and memory) required to construct a proof depend on a number of factors. Firstly, they depend on the inference rules available to the agent. In this paper we focus on inference in propositional logic; the agent can reason by cases, by exploring different branches of a logical proof. The second key factor in determining resource requirements is the architecture of the agent. We consider a sequential (Von Neumann-style) architecture, since most existing reasoners perform sequential forms of reasoning, and/or implement different forms of reasoning on sequential hardware architectures.

M. Fisher, F. Sadri, and M. Thielscher (Eds.): CLIMA IX, LNAI 5405, pp. 77–96, 2009.

We show how this behaviour can be encoded as a planning problem in two different ways, which are both interesting in that they require different planning capabilities, and provide different degrees of adaptability to representing other architectures.

In the first approach, which we call 'architecture-oriented' since it mimics the sequential computation of a proof on a Von Neumann-style architecture, we explicitly model stack manipulation. This kind of modelling gives rise to an encoding of proof search as a classical planning problem. In the second approach, which we call 'proof-oriented' since it mimics the construction of logical proofs following their tree shape, the model explores different branches of the proof in parallel; still, the model represents faithfully the memory usage by the underlying Von Neumann reasoner, including stack usage. This model gives rise to an encoding of reasoning as a strong planning problem.

We implement both models for the MBP planner[3]. Our experimental results demonstrate that even for simple problems, determining bounds is not trivial, since time and memory saving strategies may not be intuitive. Also, they allow us to experimentally compare the different features of the models in terms of planning performance.

The paper is organized as follows. First, we discuss our reference reasoning agent (Section 2). Then, we discuss the analogy between an automated reasoning problem and a planning problem (Sections 3) and different planning encodings (Sections 4 and 5). This leads to a discussion of the implementation of such encodings on the MBP planner (Section 7). Finally, we show that our approach is general enough to be easily adaptable to different models for computational resources (Section 8) and we recap with a discussion on related and future work (Section 9).

2   Reasoning with Bounded Resources

The Reasoner Model. We study proof search for an agent which contains a large set of classical propositional inference rules, which is similar in spirit to natural deduction and analytic tableaux. We have focused on propositional rather than first order proof systems because they are easier to map into propositional planning languages, and at the same time they provide typical examples of challenging reasoning patterns such as reasoning by cases. However it is possible to use our approach to encode proof search for higher order logics as a planning problem in a similar way.

Our proof system, illustrated in Figure 1, contains, for every classical connective $\neg$ (not), $\wedge$ (and), $\vee$ (or), $\rightarrow$ (implies), a set of rules. Following he approach of Natural Deduction, the rules are divided in introduction (I) and elimination (E) rules. Moreover they are formulated with respect to signed (true or false) formulas: $t : A$ means that A is true, $f : A$ means that A is false. Thus, also the rules themselves are signed: for instance I_t indicates a true Introduction rule while E_f indicates a false Elimination rule. The symbol | below the line means that two successor branches are created by the application f the rule. Thus, the rule

$$\frac{t : A \vee B}{t : A \mid t : B} \vee_{Et}$$

indicates an inference rule which starts from $A \vee B$ being true, and allows to create two successor branches: one where A is true and another where B is true.

$$\frac{t:\neg A}{f:A}\ \neg E_t \qquad \frac{f:\neg A}{t:A}\ \neg E_f \qquad \frac{t:A}{f:\neg A}\ \neg I_f \qquad \frac{f:A}{t:\neg A}\ \neg I_t$$

$$\frac{t:A\wedge B}{t:A,t:B}\ \wedge E_t \qquad \frac{t:A,t:B}{t:A\wedge B}\ \wedge I_t \qquad \frac{f:A}{f:A\wedge B}\ \wedge I1_f \qquad \frac{f:B}{f:A\wedge B}\ \wedge I2_f$$

$$\frac{f:A\wedge B}{f:A\mid f:B}\ \wedge E_f \qquad \frac{f:A\wedge B,t:A}{f:B}\ \wedge E1_f \qquad \frac{f:A\wedge B,t:B}{f:A}\ \wedge E2_f$$

$$\frac{t:A\vee B}{t:A\mid t:B}\ \vee E_t \qquad \frac{t:A}{t:A\vee B}\ \vee I1_t \qquad \frac{t:B}{t:A\vee B}\ \vee I2_t$$

$$\frac{f:A\vee B}{f:A,f:B}\ \vee E_f \qquad \frac{f:A,f:B}{f:A\vee B}\ \vee I_f \qquad \frac{f:A,t:A\vee B}{t:B}\ MT_1 \qquad \frac{f:B,t:A\vee B}{t:A}\ MT_2$$

$$\frac{t:A\to B}{f:A\mid t:B}\ \to E_t \qquad \frac{f:A}{t:A\to B}\ \to I1_t \qquad \frac{t:B}{t:A\to B}\ \to I2_t$$

$$\frac{f:A\to B}{t:A,f:B}\ \to E_f \qquad \frac{t:A,t:A\to B}{t:B}\ MP \qquad \frac{f:B,t:A\to B}{f:A}\ MT$$

$$\frac{t:A,f:A}{t:B}\ ExC \qquad \frac{}{t:A\mid f:A}\ Split$$

Fig. 1. The proof system

The system is complete for classical propositional logic,[1] and its subsets correspond to well-known deductive systems (such as analytic tableaux [11]). Proof search for such systems can be formulated as a special case of our proof system.

A derivation of A from a set of premises Γ is a tree where nodes are sets of signed formulas, the root is an empty set, each child node is added to the tree in accordance with the expansion rules of the system, and each branch has a node containing $t:A$. In addition, we require that all the formulas occurring in the conclusion of any inference rule used in the derivation are subformulas of the formulas from $\Gamma \cup \{A\}$. Since a derivation with the subformula property always exists, the system is still complete for classical logic. As an example, here is a simple derivation of $B_1 \vee B_2$ from the set of facts $A_1 \vee A_2$, $A_1 \to B_1$, $A_2 \to B_2$:

---

[1] A proof sketch: suppose $\Gamma \models \phi$ where Γ is a finite set of propositional formulas and ϕ is a propositional formula, not necessarily in the same alphabet. We show how to construct a derivation of ϕ from Γ. We start with all formulas in Γ labelled with t and apply the Split rule on ϕ, that is create two branches, one of which contains $t:\Gamma$ and $t:\phi$, and the other $t:\Gamma$ and $f:\phi$. We need to construct a tree where every branch contains $t:\phi$. We do not need to do anything with the first branch since it already contains $t:\phi$. The second branch will be a usual refutation proof; since the elimination rules of our system include all the tableaux rules from [11], the same proof as in [11] can be used to show that every branch is guaranteed to end in contradiction, and then we can derive $t:\phi$ by ExC. Note that every formula in the derivation we just constructed is a subformula of Γ or ϕ.

$$\frac{\dfrac{t:A_1 \to B_1 \quad t:A_1}{t:B_1} \, MP}{t:B_1 \vee B_2} \, \vee_{I1t} \quad \Bigg| \quad \frac{\dfrac{t:A_1 \vee A_2}{\quad} \vee_{Et}}{t:B_1 \vee B_2} \, \vee_{I2t}$$

The Architecture. We consider agents modelled as Von Neumann reasoners, which act sequentially, attempting to derive a sentence (goal formula) ϕ from a set of facts stored in a knowledge base K. We assume that K is stored outside the agent's memory, and that instead, reasoning can only take place operating on formulas stored in memory. That is, facts must be first read into the memory, and then elaborated. The agent's memory consists of m memory cells. Our working assumption is that each cell holds a single formula (i.e., we abstract away from the size of the formulas). Applying an inference rule or loading new data from K may overwrite the contents of a memory cell.

As customary in a Von Neumann-style architecture, the memory is organized into a *call stack* and a *heap* sections. The stack is used to deal with the branching of computation threads, which in our case, is due to the application of branching reasoning rules such as e.g. $\vee_{Et}$. More in detail, it is used to save and restore memory contexts at the branching points, and is a Last-In First-Out structure organized in *frames*: a frame represents a memory context saved when branching, and push/pop operations act by creating or restoring entire stack frames. The *heap* is exploited by the reasoner by reading and overwriting memory cells, accessing cells directly via their address.

The Problem Statement. To motivate the question of resource bounds, consider the derivation of $B_1 \vee B_2$ given above. Figure 2 shows one possible way in which a reasoning agent with 4 memory cells can compute the derivation. In this example the agent only manipulates true formulae, and we therefore omit the prefix t: In addition to the application of logical rules the reasoner needs to 'Read' formulae from K, can 'Overwrite' formulae, and can also 'Backtrack' if needed. In the table below, we omit the detailed description of the different overwrites as they are clear from the steps performed to produce the proof.

In step 1 the formula $A_1 \vee A_2$ is loaded into memory. At step 2 the $\vee_{Et}$ rule requires reasoning by cases; the agent decides to consider the left branch first, i.e. A_1, and to save the current state (i.e., $A_1 \vee A_2$) and other case A_2 on the stack. Steps 3–5 derive

#	Applied rule	Memory content	Stack
1	Read $(A_1 \vee A_2)$	$\{A_1 \vee A_2\}$	$\emptyset$
2	$\vee_{Et}$, branch=left	$\{A_1\}$	$\{A_1 \vee A_2, A_2\}$
3	Read $(A_1 \to B_1)$	$\{A_1, A_1 \to B_1\}$	$\{A_1 \vee A_2, A_2\}$
4	MP	$\{A_1, B_1\}$	$\{A_1 \vee A_2, A_2\}$
5	$\vee_{I1t}$	$\{A_1, B_1 \vee B_2\}$	$\{A_1 \vee A_2, A_2\}$
6	Backtrack	$\{A_1 \vee A_2, A_2\}$	$\emptyset$
7	Read $(A_2 \to B_2)$	$\{A_2, A_2 \to B_2\}$	$\emptyset$
8	MP	$\{A_2, B_2\}$	$\emptyset$
9	$\vee_{I2t}$	$\{A_2, B_1 \vee B_2\}$	$\emptyset$

Fig. 2. A proof of $B_1 \vee B_2$ (left-first)

$B_1 \lor B_2$ by first reading $A_1 \rightarrow B_1$ from K and then using Modus Ponens and or introduction. Note that while the stack cannot be overwritten, the memory cells in the heap can, and the reasoner uses this feature to limit memory usage. Now backtracking takes place, and the stack content (A_2) is popped and put in memory; steps 7–9 derive $B_1 \lor B_2$ in a manner similar to the left branch.

While Figure 2 shows a derivation of $B_1 \lor B_2$ with 4 memory cells, are these the minimal memory requirements for a proof of this formula or can we achieve it using less memory resources? And if so, can we still perform the derivation in 9 steps – as we do in Figure 2 – or is there a different trade-off between time and memory usage? These are key questions in determining the computational resources (time and memory) required to construct a proof and in the rest of the paper we show how to answer them using planning.

We take the memory requirement of a derivation to be the maximal number of formulas in memory (heap and stack) on any branch of the derivation, and the time requirement to be the total number of inference steps on all the branches in the derivation. In a system with rules of conjunction introduction and elimination, as in our system, the 'number of formulas' measure would normally give a trivial constant space complexity (since any number of formulas can be connected by 'and' until they are needed). However, in our case, the restriction of $\land_{It}$ to satisfy the subformula property prevents this from happening. For this reason, and in the interest of simplicity, we use the number of formulas measure for most of the paper. It is straightforward to replace this measure with the number of symbols required to represent the formulas in memory, which we do in section 8.

Our approach to measuring the time and space requirements of proofs is very similar to that adopted in the proof complexity literature [7,1]. Research in proof complexity considers non-branching proofs, and aims to establish, for various proof systems, the lower bounds on the length of proof and on the space required by the proof as a function of the size of the premises.

3   Bounded Resource Reasoning as Planning

To discuss the way bounded reasoning can be recast as planning, we start by a brief summary of the basic concepts and terminology in planning. A planning problem $P = (D, G)$ refers to a *domain D* and to a *goal G*. The domain $D = (F, O, A)$ represents the relevant field of discourse by means of a finite set F of *fluents* (for which initial values are defined), of a finite set O of *observations* on them, and of a finite set A of *actions* that can be performed; actions are defined in terms of *preconditions* which constrain their executability, and of *effects* which define the way they affect the values of fluents. The goal G describes a set of desirable domain configurations which ought to be attained. Then, the problem (D, G) consists in *building an entity π, called a plan, that interacts with D by reading its observations and commanding it with actions, and such that, for all possible behaviours of D when controlled by π, D satisfies G.*

Let us now consider the problem $Ex(R, K, \phi_G, M)$ of identifying the existence and minimal length of a deduction for a formula ϕ_G, starting from a set of facts K, for a reasoning agent whose memory has size M, and armed with a set of deduction rules R.

In this setting, we can consider the agent as a planning domain. E.g., its memory elements can be mapped to fluents, which are fully observable to the agent; its knowledge, i.e. the values of the fluents, changes due to its usage of reasoning rules and facts - which therefore can be perceived as the actions at stake. The goal consists in building a plan (the proof) that leads the domain to finally achieve (for all possible outcomes of deduction rules) a situation where some memory cell contains ϕ_G. Under this view, the problem $Ex(R, K, \phi_G, M)$ is recast, using a function $BuildDom$, into that of finding the shortest plan to achieve ϕ_G on a planning domain $BuildDom(R, K, M)$ whose actions correspond to the reasoning rules R and to the facts K, and whose fluents represent the memory state. This makes evident a general analogy between a bounded reasoning problem and one of planning. Of course, within this framework, different recastings, i.e. different $BuildDom$ functions, are possible and useful, since they allow representing different kind of reasoning agents and provide different ways to establish resource bounds. In turn, the different models lead to different classes of planning problems, for which different techniques are required.

A first model we consider is designed to faithfully represent the behaviour of a Von Neumann propositional reasoning agent, so to allow a direct computation of time and memory bounds for it. To appropriately mimic the execution of reasoning on such architecture, such an "architecture-oriented" model must also represent the way branching is handled by saving memory contexts on the stack, and restoring them later on. That is, in this model, multiple outcomes of branching rules are "determined", faithfully representing the way the agent decides in which order to handle threads by making use of the stack. In this setting, a solution plan is a sequence of deterministic actions (including stack manipulation), which represent the reasoning agent's execution, and whose length directly measures the time spent by the agent. This means that the problem is one of *classical* planning, and that, in particular, *optimal* classical planning algorithms allow establishing the minimal time bound for a memory-bounded deduction.

A different modelling that can be considered is "proof-oriented", in that it aims at obtaining plans whose structure is the same of (tree-shaped) proofs. To achieve this, actions must have the same structure of reasoning rules; in particular, branching rules are mapped to *nondeterministic* actions with multiple outcomes. In this setting, due to the presence of nondeterminism, the planning problem becomes one of *strong* planning [6]: the goal ϕ_G needs to be finally achieved in spite of nondeterminism. Of course, similar to the "architecture-oriented" model, also in this model the stack evolution must be kept into account, on all proof branches, to faithfully represent the way the Von Neumann reasoning agent makes use of the memory. A stack-free version of such a model can be adopted, instead, to directly represent the behaviour of a *parallel* reasoning agent where multiple processing units, each owning a private memory storage, handle different branches of the proof. Again, in this case, if *optimal* strong planning is performed, the depth of the obtained plan identifies the minimal time for deducing ϕ_G within the given memory bound.

While both modelling styles are conceptually rather simple, *effective* representations of the bounded reasoning problem in terms of planning are not immediate, and are discussed in the next two sections.

4   Bounded Reasoning by Classical Planning

In general, modeling a portion of the real-world as a planning domain must obey two different and contrasting requirements. While the domain must *faithfully* represent all relevant aspects of the problem at stake, the state-space explosion issue implies that the domain must be as compact as possible, which is usually achieved by considering *abstractions* of the actual problem. This requires finding a careful balance for the problem being examined. In the following, we start by discussing the way we model the knowledge of our reference Von Neumann reasoning agent, which is the key decision point since, then, the modeling of actions is essentially implied.

Representing Knowledge. For our problem, an immediate mapping of the agent's (heap and stack) memory cells as domain fluents, whose content represent the formulas stored in the cells, is easy to devise. However, such modeling is not effective, due to the fact that it fully represents the positional aspect of memory, i.e. *where* facts are stored. For our purposes, this is unneeded and harmful: to represent the agent's knowledge, it is irrelevant to know where a fact is stored on the heap, or where on the stack; all that is needed is to know *whether* a fact is stored on the heap, and *at which frames* it is stored on the stack. Thus we adopt a different modeling style, where, instead of having a fluent represent a cell, a fluent will represent a formula ϕ: ϕ will have an associated fluent $\mathbf{F}(\phi)$, which encodes (a) whether the formula is stored on the heap, and (b) for each stack frame, whether it is stored there.

Considering that, if memory has size M, the number of stack frames is bounded by $M - 1$ (since each frame occupies at least one cell), this means that each fluent is a string of M ternary values in $\{\top, \bot, \epsilon\}$:

$$\mathbf{F}(\phi_1) : \{\top, \bot, \epsilon\}^M \ \ldots \ \mathbf{F}(\phi_n) : \{\top, \bot, \epsilon\}^M$$

The digits of a fluent $\mathbf{F}(\phi_i)$ refer, right to left, to the heap, to the first stack frame, to the second stack frame, and so on up to the $(M - 1)^{th}$ stack frame; namely, the value of the j-th digit $\mathbf{D}_j(\mathbf{F}(\phi_i))$ represents whether ϕ_i, at the specific area associated to the digit, is known to be true ($\top$), known to be false ($\bot$), or not known (ϵ).

This model must be enriched by describing the overall memory bound, stating that *"the memory usage U, i.e. the total number of $\top$ and $\bot$ symbols in $\mathbf{F}(\phi_1), \ldots, \mathbf{F}(\phi_n)$, may never exceed M"*:

$$U = |\{(\phi_i, j) : \mathbf{D}_j(\mathbf{F}(\phi_i)) \neq \epsilon\}| \leq M \tag{1}$$

This encoding assumes that:

- reasoning only takes place considering heap values, i.e. the least significant digit (LSD - i.e. $\mathbf{D}_0(\mathbf{F}(\phi_i))$) of the fluents;
- when backtracking occurs, popping the top stack frame means simply "shifting right" each string;
- when branching occurs, in essence, pushing a new stack frame means "shifting left, retaining the LSD" for all fluents. On top of this, each formula ϕ_i generated by a branching rule must be appropriately stored in the heap or in the top stack frame, affecting $\mathbf{D}_0(\mathbf{F}(\phi_i))$ and $\mathbf{D}_1(\mathbf{F}(\phi_i))$.

Representing Actions. As mentioned, the domain actions represent the application of reasoning rules, and the usage of facts. In particular, rule schemata can be presented as planning action schemata, whose arguments represent the rule premises, and allow their instantiation onto specific formulas. E.g., the action schema $\vee_{I1t}(\psi_1, \psi_2)$, with $\psi_{1,2} \in \{\phi_1, \ldots, \phi_n\}$, indicates the application of $\vee_{I1t}$ on formula ψ_1, which is known to hold (i.e. $\mathbf{D}_0(\mathbf{F}(\psi_1)) = \top$), to derive that the formula $\psi_1 \vee \psi_2$ also holds (i.e. $\mathbf{D}_0(\mathbf{F}(\psi_1 \vee \psi_2)) = \top$). Notice that ψ_2 also appears as an argument in the schema, since it is necessary for the agent to choose explicitly which formula must be unioned to ψ_1, in order to fully instantiate the rule.

To choose amongst applicable rules and facts is not the only choice the agent must take, and such choices need to be also explicitly represented as rule arguments:

- the agent must consider the possibility of *overwriting* formulas on the heap; this is essential to save memory, and is a choice that can be taken for every outcome of the fact/rule being applied. This means that a rule that produces n outcomes in the heap comes together with n arguments $\omega_1, \ldots, \omega_n$ ranging in $\{\epsilon, \phi_1, \ldots, \phi_n\}$, specifying whether the n-th outcome must overwrite a known formula, and if so, which one. E.g., $\vee_{I1t}$ will be of the form $\vee_{I1t}(\psi_1, \psi_2, \omega_1)$.
- in the case of branching rules, the agent must decide in which order to proceed, i.e. what to keep on the heap, and what to push on the stack. In particular, for rules with two branches, such as the ones we consider, this implies adding a "left/right" direction argument δ. Then, $\vee_{Et}$ has the form $\vee_{Et}(\psi_1, \omega_1, \delta)$; notice that since only one of its outcomes stays on the heap, only one ω argument is needed.

The following rule schema defines the available actions A for the agent, which essentially consist of those in Figure 1, plus the read and backtracking actions:

$$\neg_{Et}(\psi_1, \omega_1) \qquad \neg_{Ef}(\psi_1, \omega_1) \qquad \neg_{If}(\psi_1, \omega_1) \qquad \neg_{It}(\psi_1, \omega_1)$$

$$\wedge_{Et}(\psi_1, \psi_2, \omega_1, \omega_2) \quad \wedge_{It}(\psi_1, \psi_2, \omega_1) \quad \wedge_{I1f}(\psi_1, \psi_2, \omega_1) \quad \wedge_{I2f}(\psi_1, \psi_2, \omega_1)$$

$$\wedge_{Ef}(\psi_1, \omega_1, \delta) \qquad \wedge_{E1f}(\psi_1, \omega_1) \qquad \wedge_{E2f}(\psi_1, \omega_1)$$

$$\vee_{Et}(\psi_1, \omega_1, \delta) \quad \vee_{I1t}(\psi_1, \psi_2, \omega_1) \quad \vee_{I2t}(\psi_1, \psi_2, \omega_1)$$

$$\vee_{Ef}(\psi_1, \omega_1, \omega_2) \quad \vee_{If}(\psi_1, \psi_2, \omega_1) \quad MT_1(\psi_1, \psi_2, \omega_1) \quad MT_2(\psi_1, \psi_2, \omega_1)$$

$$\rightarrow_{Et}(\psi_1, \psi_2, \omega_1) \qquad \rightarrow_{I1t}(\psi_1, \omega_1) \qquad \rightarrow_{I2t}(\psi_1, \omega_1)$$

$$\rightarrow_{Ef}(\psi_1, \omega_1, \delta) \qquad MP(\psi_1, \psi_2, \omega_1) \qquad MT(\psi_1, \psi_2, \omega_1)$$

$$ExC(\psi_1, \psi_2, \omega_1) \qquad Split(\psi_1, \omega_1, \delta) \qquad Btk \qquad Read(\psi_1, \omega_1)$$

Note that it is possible to univoquely instantiate all rules using a single ϕ argument, which, depending on the rule, can be either the premise or the result of the rule. For instance, in the case of $\vee_{I2t}$, it is possible to use a single argument ψ corresponding to $\psi_1 \vee \psi_2$, which would implicitly define ψ_1 and ψ_2 as the rule premises. While this is actually used in the implementation, here, for the sake of clarity, we keep the semantics of arguments as being premises of rules.

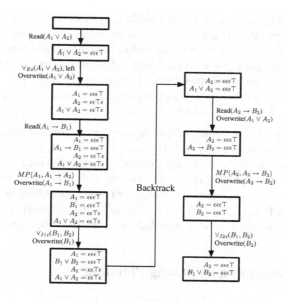

Fig. 3. A 'sequential' proof

Each rule schema comes with its specific executability preconditions, posing constraints on the current knowledge, and with effects specifying constraints on the next memory configuration. For instance, the precondition for $\vee_{I1t}(\psi_1, \psi_2, \omega_1)$ requires that ψ_1 is known to hold (i.e. $\mathbf{D}_0(\mathbf{F}(\psi_1)) = \top$). We omit the complete description of preconditions and effects for lack of space. We only note that they can be obtained almost directly from the semantics of the connectives related to the corresponding rule. On top of this, all actions have two kinds of general contraints. First, "overwrite" arguments must refer to facts actually on the heap, i.e. $\forall i : \mathbf{D}_0(\omega_i) \neq \epsilon$. Second, an action is not executable if its application would cause a memory overflow. For each action instance, this is easily decided a priori from its execution, taking into account the current memory usage U, the number of previously unknown formulas produced by the rule, and the number of overwritten formulas.

Naturally, also the result of the application of rules is specific to each rule. In particular:

- reading facts from K only affects the LSD of the fluent associated to the fact, and possibly the LSD of an overwritten formula. For instance, $Read(A1 \vee A2, \epsilon)$ only sets $\mathbf{D}_0(\mathbf{F}(A1 \vee A2)) = \top$.
- non-branching reasoning rules only affect the LSDs of some fluents: those related to the formula(s) produced by the rule application, and those related to the overwritten formulas. All other digits, and all other fluents, are unaffected. For instance, $\vee_{I1t}(B_1, B_2, B_2)$, which produces $B_1 \vee B_2$ overwriting it in place of B_2, affects only $\mathbf{D}_0(\mathbf{F}(B_1 \vee B_2))$, which becomes $\top$, and $\mathbf{D}_0(\mathbf{F}(B_2))$, which becomes ϵ.
- popping a stack frame, i.e., in terms of reasoning, backtracking, simply shifts right all fluent strings, introducing an ϵ at their most significant digit.

- branching rules operate on all fluents, by shifting them left and keeping their LSD; moreover, the two fluents correspondent to the rule outcomes are affected so that in one, the LSD is set to a truth value, and in the other, the second digit is set to a truth value - representing that one formula is dealt immediately on the heap, while the other is stored at the top stack frame. For instance, starting from $\mathbf{F}(A_1 \vee A_2) = \epsilon\top$ and $\mathbf{F}(A_1) = \mathbf{F}(A_2) = \epsilon\epsilon$, applying $\vee_{Et}(A_1 \vee A_2, A_1 \vee A_2, left)$ leads to $\mathbf{F}(A_1 \vee A_2) = \mathbf{F}(A_2) = \top\epsilon$ and $\mathbf{F}(A_1) = \epsilon\top$.

In Figure 3 we show an example of a deduction for our running example, where the facts in the knowledge base are $K = \{A_1 \vee A_2, A_1 \to B_1, A_2 \to B_2\}$ and the goal is $B_1 \vee B_2$. We consider an agent with 4 memory cells, and for the sake of simplicity, from the set of fluents associated to $\{A_1, A_2, B_1, B_2, A_1 \vee A_2, B_1 \vee B_2, A_1 \to B_1, A_2 \to B_2\}$, we omit showing those whose formula is not stored anywhere in memory, that is fluents of the form $\mathbf{F}(\phi) = \epsilon\epsilon\epsilon\epsilon$. Also, for the sake of readability, we leave the $\mathbf{F}(\cdot)$ notation implicit. The proof shows that a deduction for $B_1 \vee B_2$ is possible in 4 cells, taking 9 steps.

5 Bounded Reasoning by Strong Planning

While the previous model is "architecture-oriented", one can follow a "proof-oriented" modelling where the produced proofs are shaped as logical trees. As we shall see, one advantage of such a model is that it can further abstract from the specific memory configuration, while still representing the relevant features of our reference reasoning agent. Moreover, such a model can be easily adapted to represent reasoning agents based on a parallel architecture. Again, we will show this by first discussing the way the agent's knowledge is represented.

Representing Knowledge. In a "proof-oriented" model, different than in the previous "architecture-oriented" model, separate proof branches correspond to deduction threads evolving in parallel, independently. This will be mapped into the planning process; hence, in such a model, proof search can be performed with no need to represent any context push/pop mechanism. Nevertheless, at each proof step, in order to evaluate memory consumption, we need to represent the actual memory usage of the underlying Von Neumann reasoning agent, which actually uses the stack. Then, to achieve a representation of knowledge as compact as possible, we build on top of the "position abstraction" taken for the previous model, and apply a further abstraction, so that:

1. concerning formulas, we only represent the *current* knowledge on them, i.e. their heap values;
2. concerning the stack, we only represent its current size.

This leads to a modelling where formula-representing fluents are 3-valued, and an additional numeric fluent S represents the stack size:

$$\mathbf{F}(\phi_1) \in \{\top, \bot, \epsilon\}$$

$$\cdots$$

$$\mathbf{F}(\phi_n) \in \{\top, \bot, \epsilon\}$$
$$S \in [0, M-1]$$

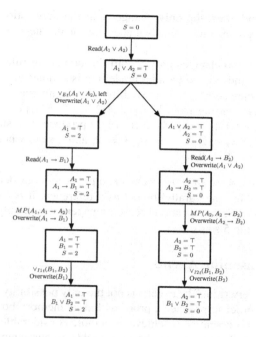

Fig. 4. A 'branching' proof

Again, the bound on the memory usage U must be added; of course here, U is computed in a different way, although it produces the same result:

$$U = S + |\phi_i : \phi_i \neq \epsilon| \leq M \tag{2}$$

It is immediate to see that such a representation is significantly smaller, in terms of possible states, than the one considered in the previous model, since different stack configurations with equivalent memory occupation are grouped together. On the other side, this representation features, together with an additional state-level branching factor implied by the fact that rules correspond to nondeterministic actions, transform the problem from classical to strong planning.

It is important to notice that a stack-free version of this modelling provides a direct representation of a different, parallel reasoner architecture, where reasoning is actually performed in parallel by different processing units, each featuring a private memory bank of size M. In this case, the plan depth directly represents the overall time spent by the parallel architecture to complete the proof, and optimal planning can be used to establish a minimal bound in that respect.

Representing Actions. In terms of available actions and arguments, the parallel modelling is substantially the same as the sequential one, apart from the absence of the 'backtracking' operator. Of course, what changes is the way actions are defined, in terms of preconditions and effects.

In terms of preconditions, the only change is in the description of the overflow-avoiding executability constraint, which makes use of the new computation for the memory usage U.

In terms of effects, the changes only regard the branching rules, and in that case they are considerable. Indeed, the parallel modelling is simpler in that, when two outcomes are produced, they correspond to two different assignments to the sets of fluents; therefore, for instance, starting from $\mathbf{F}(A_1 \vee A_2) = \top$, $\mathbf{F}(A_1) = \mathbf{F}(A_2) = \epsilon$, and $S = 0$, and applying $\vee_{Et}(A_1 \vee A_2, A_1 \vee A_2, left)$ produces two states: a state where $\mathbf{F}(A_1 \vee A_2) = \mathbf{F}(A_2) = \epsilon$, $\mathbf{F}(A_1) = \top$, $S = 2$, and a state where $\mathbf{F}(A_1 \vee A_2) = \mathbf{F}(A_1) = \top$, $\mathbf{F}(A_2) = \epsilon$, $S = 0$.

An example. In Figure 4 we show the same example proof considered in the previous section model, but taking the parallel model as a reference. It is easy to see, by discarding the stack fluent S, that a parallel reasoning agent would only require 2 cells to complete such a proof in 5 steps.

6   Forgetful Reasoning Agents

The possibility of "overwriting" heap cells is not the only possibility that an agent must take into account in order to produce a proof within its memory bounds. Indeed, considering other options is essential to achieve even more considerable memory savings.

In particular, as a skilled reader may have spotted from our example proofs already, a key issue is related to the behaviour in the presence of branching rules. There, the standard behaviour implemented by a Von Neumann machine stores the whole current context on the stack. However, it may well be the case that some of the facts currently known will not be useful in the pending branch, and therefore, pushing them on the stack only wastes memory.

A "smart" agent would act by selecting which "relevant" facts must be stored on the stack, and storing only them. For instance, in our example, $A_1 \vee A_2$ is not used in the second branch of the proof; and even if it were, since it is an axiom, it could have been re-read, rather than saved in memory. So a "smart" agent would have been "forgetful" w.r.t. saving $A_1 \vee A_2$ on the stack, leading to a proof with the same schema of the one we presented, but where (a) in the left branch we would have $S = 1$, and (b) on top of the right branch, $A_1 \vee A_2 = \top$ would not appear. It is easy to see that such a proof only requires 3 cells, and still takes 9 steps.

Two remarks are in order. First, the issue of "forgetfulness" is indeed specific of stack treatment, since the ability to overwrite facts on the heap renders harmless the presence of useless facts in that memory area. Second, while, during a specific proof search, forgetting some facts may lead to not derive a fact which would be otherwise derivable, it is clear that an agent which considers forgetfulness as an option, and tries out all the different choices (including "not forgetting anything") is still complete. Note also, that here we are not concerned on whether "forgetful" agents can be realised in efficient manners. Our main point here is that "forgetfulness" is a strategy that can help to build proofs which use a smaller number of memory cells.

Given this, the agent's forgetfulness choices are modelled as additional arguments to branching rules. Namely, branching rules are enriched with additional arguments

$\rho_1, \ldots, \rho_M$ in ϵ, $\phi_1, \ldots, \phi_n$, and formula ϕ_j must be "forgotten" on the pending branch iff $\exists i : \rho_i = \phi_j$. Of course, it must be possible only to forget facts which are currently in the heap. Moreover, to prune out equivalent symmetric specifications of the same forgetfulness, we impose the following ordering constraints on the ρ_i arguments:

1. if k formulas must be forgotten, they are specified within $\rho_1, \ldots, \rho_k$, i.e. $\rho_i = \epsilon \rightarrow \forall j > i : \rho_j = \epsilon$
2. the order of forgotten formulas is reflected in the order they are specified as ρ_i arguments: $\forall i, j : \rho_i = \phi_l \wedge \rho_j = \phi_m \wedge (i < j) \rightarrow (l < m)$.

7 Implementation in MBP and Experiments

We implemented our approach using the MBP planner, devising an automated modeller which converts a bounded-resource reasoning problem (given as a set of facts, a goal formula ϕ_G and a memory bound) into a planning problem description, following one amongst our proposed models. Then, we use MBP to identify (a) whether a deduction exists for ϕ_G within the given memory bound, and (b) if so, which is the minimal number of computation steps it takes. Of course, as a side-effect, if a deduction exists, we also obtain an optimal strategy, which can be outputted or simulated.

Our choice of MBP has several reasons. First, MBP is a flexible planning tool, which combines different planning algorithms, amongst which the optimal classical and strong planning we need. Second, it is an effective tool; due to its internal data representations which are based on Binary Decision Diagrams, it has proved to be state of the art, especially for what concerns optimal planning in the presence of nondeterminism. Third, its input language, SMV, is very flexible and apt to express the kind of models at hand. In particular, the ability of SMV to express the behaviour of a domain by modularly specifying the dynamics of each of its fluents is very useful, and can be combined with the usage of constraints spanning across fluents to achieve very compact encodings – even when modelling imply nondeterministic dynamics. In particular, in our models, the dynamics of each fluent is defined independently, by a case analysis, using SMV's `assign` and `case` constructs. Each case identifies one action that affects the fluent, and the "default" case is used to specify the inertial behaviour. This allows for a very compact encoding: in no case, the SMV model exceeds 52 Kbyte of size.

Based on our automatically generated SMV encodings, we ran tests with MBP, considering 8 different deduction schemata, and 3 reference architectures: a standard Von Neumann reasoning agent, a Von Neumann reasoning agent considering forgetfulness, and a parallel reasoning agent. All tests have been performed with a cut-off time at 1800 seconds on a Linux machine running at 2.33GHz with 1.8Gb of RAM; results are reported in Figures 5 and 6.[2]

Problems vary in size and nature, with up to 14 fluents and 30 actions. Problem P_8 is our reference example; as one can see, in that case, one can prove $B_1 \vee B_2$ using just 2 cells, which seems surprising. The proof is reported in Figure 7, and makes use of two main ideas: (a) branching immediately, so that context saving is not required, and (b)

---

[2] The benchmark instances and executables of the planner and the modelers are available at http://boundedreasoning.i8.com/

	Von Neumann Std		Von Neumann Fgt		Parallel	
	min. M	min. t	min. M	min. t	min. M	min. t
P_1	4 $(t \geq 10)$	7 $(M \geq 5)$	4 $(t \geq 7)$	7 $(M \geq 4)$	3 $(t \geq 5)$	5 $(M \geq 3)$
P_2	3 $(t \geq 11)$	10 $(M \geq 4)$	3 $(t \geq 11)$	10 $(M \geq 4)$	2 $(t \geq 9)$	6 $(M \geq 3)$
P_3	4 $(t \geq 24)$	24 $(M \geq 4)$	4 $(t \geq 24)$	24 $(M \geq 4)$	3 $(t \geq 9)$	9 $(M \geq 3)$
P_4	3 $(t \geq 13)$	13 $(M \geq 3)$	3 $(t \geq 13)$	13 $(M \geq 3)$	3 $(t \geq 6)$	6 $(M \geq 3)$
P_5	3 $(t \geq 9)$	9 $(M \geq 3)$	3 $(t \geq 9)$	9 $(M \geq 3)$	3 $(t \geq 5)$	5 $(M \geq 3)$
P_6	3 $(t \geq 16)$	13 $(M \geq 4)$	3 $(t \geq 15)$	13 $(M \geq 4)$	2 $(t \geq 8)$	7 $M \geq 3)$
P_7	3 $(t \geq 8)$	8 $(M \geq 3)$	3 $(t \geq 8)$	8 $(M \geq 3)$	2 $(t \geq 4)$	4 $(M \geq 2)$
P_8	2 $(t \geq 8)$	8 $(M \geq 2)$	2 $(t \geq 8)$	8 $(M \geq 2)$	2 $(t \geq 5)$	5 $(M \geq 2)$

Fig. 5. Memory and time bounds for the three models

	Von Neumann Std	
	standard	proof
P_1	1.20	0.27
P_2	3.28	0.61
P_3	33.43	1.82
P_4	26.25	3.46
P_5	1.83	0.33
P_6	30.14	2.83
P_7	0.79	0.24
P_8	0.24	0.10

Fig. 6. Computation times (s)

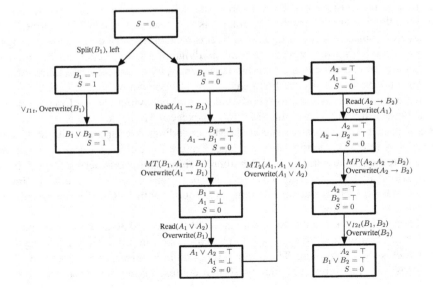

Fig. 7. A proof with 2 cells

carefully selecting the order in which branches are handled, so that "the first branches are the easier ones to solve" (since they have some memory occupied by the stack). As a result, the proof is very different from the intuitive ones shown in the previous sections. The fact that bounds are not trivial, and neither are proof-saving strategies, is also true for most of the other examples.

More in general, Figure 5 shows an evident trade-off between space and time: when minimal memory is used (denoted by 'min. M' in the results), more time is required, and faster proofs require more memory (we denote it by 'min. t', giving in parenthesis the memory used for that minimum proof, within the memory and time cutoff used). Also, the experimental results show clearly the importance of forgetfulness to save memory: in problems P_1 and P_4, the minimal bounds of memory for a forgetful Von Neumann reasoning agent improve those of a standard reasoning agent (while not requiring more time).

In terms of MBP performance, it is interesting to compare the relative performances of the "classical" and "strong" models of the problem for Von Neumann reasoning agents. Results for the forgetful Von Neumann reasoning agent are presented in Figure 6, and show that there is a clear advantage of the "proof-oriented" model. This indicates that the effective way MBP handles nondeterminism pays off, and hints at using the "strong" first to identify memory bounds, leaving the "classical" model to check time limits, once the memory bound is fixed.

8 Different Resource Models

While our work took definite assumptions over the underlying model for computational resources, established in Section 3, our approach is general enough to be easily adaptable to different views. In particular, different data storage means and methodologies can be mapped into different, but equally interesting, memory bound models. For instance, taking the assumption that every formula has size one is very apt to represent situations where the actual representation of formulae is not stored in the working memory, but on an external mass storage, and the data on the heap and stack are simply "pointers" to such representations. Vice versa, the memory usage of a reasoning system where formulae are stored in the main memory cannot be faithfully captured by such a model, since in that case, different formulae occupy different memory quotas. In this case, we can take the reasonable assumption that a formula is represented as a parse tree, where logical connectives occupy single memory cells, as well as atomic facts. Thus, for instance, the formula $(A \land B) \lor C$ occupies 5 memory cells.

We now show that shifting from the previous, "unitary formula size" model to this new "structural formula size" model is very simple to achieve. In fact, all we need to affect is the way the memory bound is described, by introducing a function $\mathbf{S}(\phi_i)$ that associates a fact ϕ_i to a positive integer representing its memory occupation. Considering the modeling described in Section 4 to solve the problem by classical planning, this means that the overall memory bound described by Equation (1) must become

$$U = \sum_i \sum_j (\mathbf{S}(\phi_i) \times \mathbf{A}(\phi_i, j)) \leq M$$

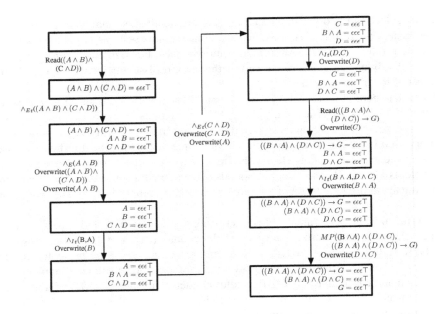

Fig. 8. P_9: a derivation of G in 9 steps

where $\mathbf{A}(\phi_i, j)$ attains value 1 if $\mathbf{D}_j(\mathbf{F}(\phi_i)) \neq \epsilon$, and attains value 0 otherwise. Similarly, in the modeling described in Section 5 to solve the problem by strong planning, the constraint described by Equation (2) must be replaced by

$$U = S + \sum_i (\mathbf{S}(\phi_i) \times \mathbf{A}(\phi_i)) \leq M$$

where $\mathbf{A}(\phi_i)$, this time, has value 1 if $\phi_i \neq \epsilon$, and value 0 otherwise. Note that, in both cases, replacing the $\mathbf{S}$ function with the constant function 1 causes the modelings to fall back in the unitary formula size case.

Expressiveness considerations apart, the relevance of being able to apply different models of resource utilization stands in the fact that the optimality of derivation strategies is only relative to a specific model. We show this considering a simple scenario where the knowledge base K contains the five formulae

$$(A \wedge B) \wedge (C \wedge D) \qquad ((B \wedge A) \wedge (D \wedge C)) \to G \qquad (B \wedge A) \to G_1$$
$$(D \wedge C) \to G_2 \qquad (G_1 \wedge G_2) \to G$$

and the goal is G.

If we consider the unitary formula size model, we can run MBP over the classical model representation of the problem and we obtain that at least 3 memory cells are needed to derive G, and the corresponding shorter proof P_9 takes 9 steps. The proof is shown in Figure 8 and makes use only of the first two facts listed in the knowledge base above. No shorter proof can be found, even when more than 3 memory cells are available.

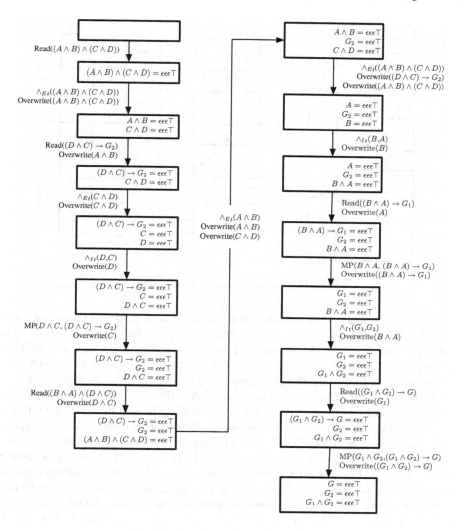

Fig. 9. P_{15}: a derivation of G in 15 steps

However, the situation is quite different when using the structural formula size model. In this case, again running MBP over the classical planning model, we obtain that the most memory-saving proof is a proof P_{15} that, making use of all the facts in K apart from $((B \wedge A) \wedge (D \wedge C)) \rightarrow G$, takes 15 steps, using 9 memory cells (see Figure 9).

Intuitively, in this model, applying the modus ponens rule over the facts $(B \wedge A) \wedge (D \wedge C)$ and $((B \wedge A) \wedge (D \wedge C)) \rightarrow G$, like P_9 does, is too memory-expensive, as these two formulas, together, occupy 16 memory cells. Rather, it becomes convenient, in terms of memory usage, to go through the derivation of G_1 and G_2, which exploits "smaller" facts (but takes additional derivation steps). This consideration of course did

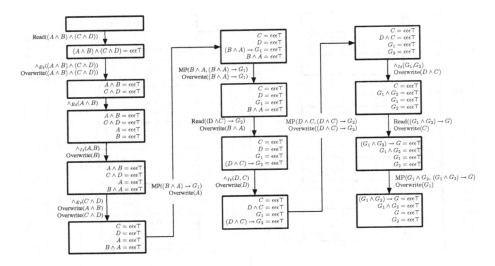

Fig. 10. P_{13}: a derivation of G in 13 steps

not apply to the unitary size model, where all formulae are regarded as having the same size. We also observe that in P_{15}, again to save memory, the fact $(A \wedge B) \wedge (C \wedge D)$ is read twice from K (since it is needed twice, but it is overwritten). Indeed, if some more memory is available (between 10 and 15 cells), the proof P_{13} is generated, which does not perform such re-reading, therefore saving two reasoning steps (see Figure 10).

Finally, if at least 16 memory cells are available, the proof P_9 is identified as the faster way to obtain G (in 9 steps). These results indicate that, in the structural formula size model, P_{15}, P_{13}, P_9 are all interesting, as they express different memory-space trade-offs: each of them takes longer, but is more memory-efficient, than the next. Vice versa, in the unitary formula size model, P_{15} and P_{13} are not more memory-effective than P_9 - actually, P_{13} is even more memory consuming. As such, in the unitary formula size model there is no reason for using P_{15} or P_{13}, under any possible resource configuration. Of course, these considerations about the dependency of optimality trade-offs on the resource models are general, and different memory models can be very easily accommodated by appropriately devising the **S** function.

9   Conclusions and Future Work

The study of the behaviour of systems in the presence of computational resource bounds is receiving considerable attention, due to its practical impacts. This has led to developing models and approaches to deal with bounded rationality, with a particular emphasis on the design of anytime algorithms to work within strict time limits (see e.g. [10]).

In this work, we link reasoning to planning, and provide an approach to evaluate the minimal time and memory bounds for solving reasoning problems, a crucial step to appropriately dimension the computational power of devices that support reasoning functionalities. While the connection between deduction and planning has already been established for a variety of logics, e.g., temporal, linear and propositional logics, see [8,4,9], existing work has essentially focused on using theorem provers to build plans. Here, we go the other way around, showing that reasoning - and taking into account resource bounds - can be recast in terms of planning, and that different re-castings are possible, interesting and useful. Our experiments show that, already considering simple scenarios, interesting memory-space trade-offs appear; minimal resource bounds are often surprising, and so are the reasoning strategies enforcing them.

While the focus of this work is on propositional logics, it has to be remarked that the general approach is open to extensions to different logics and logic fragments, which can be pre-compiled into planning domains following the conceptual scheme presented here. Indeed, a further direction of investigation, which we intend to pursue, concerns the relation between the power of the reasoning system, and the minimal resource bounds for performing deductions. Studying this can help identifying the proper 'kind' of reasoning agents, given the resource bounds and the kind of proofs that need be carried out. Also, our approach can be very easily adapted to follow different memory models, e.g. one where the occupancy of formulas depends on their internal structure. Such different models may imply different bounds, and different time- and memory-saving strategies: for instance, in our model, conjunction introduction can be used to compress formulas to save space (trading off time for space); in different models, this may not be the case.

A further relevant concern stands in improving the scalability of our approach, to cover complex scenarios where reasoning taks place over large sets of facts. This is crucial to lift the approach to practical settings, using deduction to model the behavior of actual bounded-reasoning agents. At the current stage, however, scalability is limited, since we perform an exhaustive blind plan search, in order to never rule out any resource-optimal deduction. This makes the complexity of identifying optimal resource bounds extremely high, and in such a setting even the most effective technologies for representing and searching proof trees can only help to a limited extent. To solve this issue, we need to identify heuristics or strategies that significantly reduce the complexity of the search, adapting planning techniques which embed resource optimization, such as those of [5], or techniques which allow describing control rules, such as those of [2]. Of course, the critical issue in this respect stands in guaranteeing that such heuristics and strategies preserve resource optimality. This is far from trivial, since, as witnessed in our examples, resource-optimal strategies can be far from intuitive. Such a challenging problem is the key long term goal of our research line.

References

1. Alekhnovich, M., Ben-Sasson, E., Razborov, A.A., Wigderson, A.: Space complexity in propositional calculus. SIAM J. Comput. 31(4), 1184–1211 (2002)
2. Bacchus, F., Kabanza, F.: Using Temporal Logics to Express Search Control Knowledge for Planning. Artificial Intelligence 116(1-2), 123–191 (2000)
3. Bertoli, P., Cimatti, A., Pistore, M., Roveri, M., Traverso, P.: MBP: a model based planner. In: Proceedings of the IJCAI 2001 Workshop on Planning under Uncertainty and Incomplete Information, Seattle (August 2001)

A Formal Framework for User Centric Control of Probabilistic Multi-agent Cyber-Physical Systems

Marius C. Bujorianu, Manuela L. Bujorianu, and Howard Barringer

Centre for Interdisciplinary Computational and Dynamical Analysis,
University of Manchester,UK
{Marius,Manuela.Bujorianu}@manchester.ac.uk,
Howard.Barringer@manchester.ac.uk

Abstract. Cyber physical systems are examples of a new emerging modelling paradigm that can be defined as multi-dimensional system co-engineering (MScE). In MScE, different aspects of complex systems are considered altogether, producing emergent properties, or loosing some useful ones. This holistic approach requires interdisciplinary methods that result from formal mathematical and AI co-engineering. In this paper, we propose a formal framework consisting of a reference model for multi-agent cyber physical systems, and a formal logic for expressing safety properties. The agents we consider are enabled with continuous physical mobility and evolve in an uncertain physical environment. Moreover, the model is user centric, by defining a complex control that considers the output of a runtime verification process, and possible commands of a human controller. The formal logic, called safety analysis logic (SafAL), combines probabilities with epistemic operators. In SafAL, one can specify the reachability properties of one agent, as well as prescriptive commands to the user. We define symmetry reduction semantics and a new concept of bisimulation for agents. A full abstraction theorem is presented, and it is proved that SafAL represents a logical characterization of bisimulation. A foundational study is carried out for model checking SafAL formulae against Markov models. A fundamental result states that the bisimulation preserves the probabilities of the reachable state sets.

Keywords: multi agent systems, cyber-physical systems, user centric control, stochastic model checking, bisimulation, runtime analysis, symmetries.

1 Introduction

Cyber physical systems (CPS) are tight integrations of computation with physical processes. Examples can be found in diverse areas as aerospace, automotive, chemical processes, civil infrastructure, energy, manufacturing, transportation and healthcare. A realistic formal model for CPS will consider the randomness

M. Fisher, F. Sadri, and M. Thielscher (Eds.): CLIMA IX, LNAI 5405, pp. 97–116, 2009.
© Springer-Verlag Berlin Heidelberg 2009

of the environments where there are deployed, as well as the fact that in most applications these are interacting agents. The agents are used to model entities enabled with physical mobility (e.g. cars, planes or satellites), which are able to do autonomous or guarded transitions, which are able to communicate and with evolutions continuous in both time and space. In the past all these essential system features were studied separately or shallowly integrated. The new technologies like CPS and ubiquitous computing require deep integration of orthogonal multiple features, raising the issue of modelling of emerging or lost system properties. This problem is approached in this problem by proposing a formal framework called *multi-dimensional system co-engineering* [7,6] (MScE), a holistic view combining formal, mathematical and control engineering. With this respect, formal methods for the specification and verification have been only recently developed, like *Hilbertean formal methods* [4,5] or *stochastic model checking* [10] . The most effective verification method has proved to be model checking [17]. This paper is a *foundational* study of model checking for a stochastic model of agents. In this model, each agent is a new computational model called *agent stochastic cyber-physical system (aCPS)*, i.e. each agent can move physically, thus it can be thought of as a hybrid system. Moreover, uncertainty is considered for both environment and for agent's hybrid behavior, and this uncertainty is quantified probabilistically. In the *multi-agent stochastic cyber-physical system (MAPS)* model all agents are embedded in a common physical environment and they communicate using channels. The new model, which is the kernel of MScE, addresses three new issues:

- provide real time information about the changing environment of agent based CPS.
- represent the information collected during runtime system analysis
- model the co-existence of the human control and automated control (making the model *user-centric*)

User enabled control is very important for CPS, where failures or incorrect usage may be catastrophic.

The MScE framework comprises:

- The holistic, mathematical models of aCPS and MAPS, and
- A formal logic, called *safety analysis logic* (SafAL), that offers original specification techniques for the probabilistic properties of single agent reachability. It also contains coloring types and two imperative operators: one of control theoretic nature saying that a discrete transition (a control) will take place, and a recommendation operator that prescribes a discrete transition.
- A verification strategy of safety properties expressed in SafAL against aCPS using system symmetries. We investigate the issues of bisimulation and of model checking.

There are two key concepts in model checking: *reachability analysis* and *bisimulation*. The first concept gives the effective behavior of the system, while bisimulation means the elimination of the computationally irrelevant states (duplicate or the unreachable states). Our approach departs by introducing a new and natural concept of bisimulation. Two continuous stochastic processes are considered bisimilar if their reachable sets have the same hitting probabilities.

A fully abstract semantics is constructed for the SafAL and it is proved that two states are bisimilar iff they are spatially symmetric. This result shows that bisimulation is a concept too strong for practical verification and it justifies the coloring approach. Using colors, more flexible equivalence concepts can be introduced.

Recent advances in probabilistic model checking have been achieved using the state space symmetries [19]. We use space symmetries to define a new semantics for SafAL. One main advantage of this new semantics is that we can refine the bisimulation concept. In practice, the probabilities are approximated and their equality is difficult to check. Most of current approaches consider a metric and ask that the transition probabilities differ very little [11]. In our approach, we ask the equality only for the reach set probabilities associated to some sets selected using the symmetries. One advantage of this definition is that some transition probabilities might be different, but these differences should 'compensate' when we consider global behaviors.

Using state space symmetries, we establish two important results. One of them assures the full abstraction of this new logic. The second one opens the possibility of model checking of SafAL formulas.

The paper is organized as follows. The next section succinctly presents a communicating multi-agent model. In section 3, safety analysis logic is introduced for the specification of safety properties for the multi-agent model. Section 4 is devoted to the development of a formal semantics for the logic, based on symmetries, which makes possible the model checking. In section 5, a new bisimulation concept is introduced, and a full abstraction result is proved. The paper ends with some concluding remarks. An appendix contains background material on stochastic processes.

2   A Stochastic Model for Multi-agent Cyber-Physical Systems

A cyber-physical system (CPS) is a system featuring a tight combination of, and coordination between, the system's computational and physical elements [24]. The US National Science Foundation (NSF) has identified cyber-physical systems as a key area of research. Starting in late 2006, the NSF and other United States federal agencies sponsored several workshops on cyber-physical systems.[1]

2.1   An Informal Presentation

An *agent stochastic cyber-physical system (aCPS)* is based on the concept of stochastic hybrid system [8,23] (SHS). A hybrid automaton consists of a

---

[1] "Cyber-physical systems". Grant opportunities. Grants.gov, a United States governmental resource. http://www.grants.gov/search/search.do?

discrete controller that switches between different continuous dynamical systems via some control laws modeled as guards. The evolution of every dynamical system is depicted as an open set in the Euclidean space, called location. The controller is represented as guarded transitions between modes.

The aCPS model considers nondeterminism in mode change and introduces two classes of probabilities:

• the noise in the modes; the evolution of each dynamical system is governed by a stochastic differential equation.
• the probabilities of discrete transition, described by *reset maps*. These probabilities (formally *probability kernels*) evaluate the chances to restart a given dynamical system and depend on time and on the current evolution in a mode.

The agents can have *physical mobility*, i.e. the differential equation in modes may describe the moving equations of some devices (perhaps cars, planes or satellites) in a physical environment. This mobility can be affected by the *environment uncertainty* (captured formally by the SDEs from modes) and the agent decisions (the reset maps) can be also *unpredictable* (captured by the probability kernels).

Moreover, a MAPS has communication labels, and the transitions are classified in *send* transitions and *receive* transitions. A multi-agent model then considers several SHS, each one with an associated agent, and the communications between them is done via a given set of communication channels.

To every channel l there are associated two distinct labels:

• l^s denotes the action of sending a message through the channel l, and
• l^r denotes the action of receiving a message through the channel l.

In the standard models of hybrid systems, the transitions are guarded, with guards interpreted as boundaries of the location's invariant set. In the colored version, the boundary is extended to a colored region (red in this example), where a discrete transition must be triggered.

The first extension of the SHS model which allows to model agents is given by the concept of user triggered transitions. These are unguarded transitions (i.e. mode changes) that offer to an agent more freedom in controlling their evolution. An autonomous transition can be triggered, for example, by the inter-agent communication or by a driver after being warned by the brake assistance system. In the colored model, there is a colored region (yellow in this case) where the user *is required* to perform a discrete transition.

A further development of the model requires that the two types of discrete transitions and colored region coexist. Moreover, the red regions are always included in the yellow regions, which means that a controlled transition takes place only if the user has not triggered any discrete transition. It is also essential that any of the user triggered or the controlled transition produces the same post location (otherwise it means that the user is allowed to make unsafe procedures). These conditions are depicted in Figure 2.

The models illustrated in Figures 1 and 2 can be interpreted as two different viewpoints on a CPS. In Figure 1, it is expressed the viewpoint of observing an

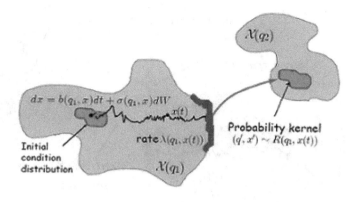

Fig. 1.

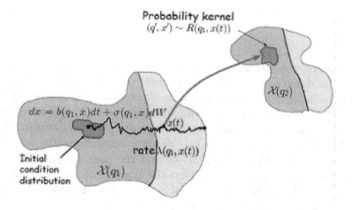

Fig. 2.

autonomous car. The viewpoint from Figure 2 is that of observing the behaviour of a driver. The model in Figure 3 can be thought of as a viewpoint integration, where the automated control and the human control coexist. Moreover, the model is user centric because the two forms of control are hierarchical, user decisions having the higher priority.

2.2   The Formal Model

An *agent* *stochastic* *cyber-physical* *system* *(aCPS)* *is* *a* *collection* $\langle (Q, d, \mathcal{X}), (y, r), ((J, \lambda, R), (m, f, \sigma)), L \rangle$ where:

(i) $(Q, d, \mathcal{X})$ describes the *state space*, which is a countable union of open sets from an euclidean space (the *modes*), each one corresponding to a discrete location. Note that the dimension of embedding euclidean space might be different for different locations.

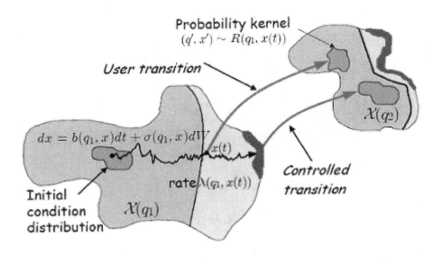

Fig. 3.

(ii) *(y,r)* is the *coloring structure*. At each time, the coloring functions give, for each mode, the dangerous region (the *guard*), colored as red, and, respectively, the potentially dangerous zone (the *safety awareness guard*) colored as yellow.

(iii) $((J, \lambda, R), (m, f, \sigma))$ gives the *transition structure*, comprising the *discrete transition structure* (J, λ, R) and the *continuous (physical) transition structure* (m, f, σ). $J = C \cup U$ is the set of all *discrete transitions (or jumps)*. This is the union of the *controlled transitions* C and the *user triggered transitions* U. The controlled transitions C depend on the transition-choice function R. λ is the *jump rate* (it determines the rate of process discrete transitions). (m, f, σ) characterizes the continuous dynamics within the modes.

(iv) L is the set of *communication channels* (or *labels*).

These entities are formally defined next.

The complex structure of the state space (i) is defined as follows:

- Q is a finite set (of *locations*),
- $d : Q \to \mathbb{N}$ is a map giving for each location the dimension of the continuous state space in that location,
- $\mathcal{X} : Q \to \mathbb{R}^{d(\cdot)}$ is the mode definition function that maps each $q \in Q$ into an open subset $\mathcal{X}(q) = X^q$ of $\mathbb{R}^{d(q)}$ i.e. for each $q \in Q$, X^q is the *mode* (the *invariant* set) associated to the location q.

Let us denote by X the whole space, i.e. $X = \cup\{(q, X^q)|q \in Q\}$. Define the boundary set $\partial X^q := \overline{X^q} \backslash X^q$ of X^q and the whole space boundary $\partial X = \cup\{(q, \partial X^q)|q \in Q\}$.

The coloring structure (ii) is given by

- two coloring functions, the yellow function $y : \mathbb{R} \to 2^X$ and the red function $r : \mathbb{R} \to 2^X$, where $\mathbb{R}$ is the set of reals. For any location, every red colored state set must be included in the yellow colored one, i.e. if $r(t) = (q, A)$ and $y(t) = (q, B)$ then $A \subset B$. Moreover, every colored set must be included in a single location, i.e. $A, B \subset \mathbb{R}^{d(q)}$.

The jump structure from (iii) is given as follows:

- Each controlled transition $v \in C$ is a quadruple $v =: (q, l, q', R_b)$ where q is the origin location, l is the label of the jump, q' is the target location, and R_v is the *reset map* of the jump (or the *green coloring* function), i.e. for each $x \in \partial X^q$ with $R(v, q, x) > 0$ (see next item) and for all Borel sets A of $X^{q'}$ the quantity $R_v(x, A)$ is the probability to jump in the set A when the transition v is taken from the guard state x (boundary state).
- U is the set of *user triggered transitions*. Each element $u \in U$ is a pentuple $u =: (q, l, q', R_u, \lambda)$, where q is the origin location, l is the label of the jump, q' is the target location, R_u is the reset map of the jump.
- The function $R : C \times Q \times \partial X \to [0, 1]$ is defined such that for all $q \in Q$, all $x \in \partial X^q$, and all $v \in C$, which are outgoing transitions of q, the quantity $R(v, q, x)$ is the probability of executing a controlled transition v. In rest, R takes the zero value. Moreover, $\sum_{v \in C_{q \to}} R(v, l, x) = 1$ for all l, x, where $C_{q \to}$ is the set of all elements of C that are outgoing transitions of q.

The continuous motion parameters from (iii) are given as follows:

- $f : X \to \mathbb{R}^{d(\cdot)}$ is a vector field
- $m : Q \to \mathbb{N}$ is a function that returns the dimension of the Wiener processes (that governs the evolution in the continuous state space, see the next item)
- $\sigma : X \to \mathbb{R}^{d(\cdot) \times m(\cdot)}$ is a $X^{(\cdot)}$-valued matrix. For all $q \in Q$, the functions $f^q : X^q \to \mathbb{R}^{d(q)}$ and $\sigma^q : X^q \to \mathbb{R}^{d(q) \times m(q)}$ are bounded and Lipschitz continuous and the continuous motions is governed by the following stochastic differential equation (SDE):

$$dx(t) = f^q(x(t))dt + \sigma^q(x(t))dW_t$$

where $(W_t, t \geq 0)$ is an $m(q)$-dimensional standard Wiener process in a complete probability space.
- Moreover, we assume the following *axioms*:
  Assumption about the diffusion coefficients: for any $i \in Q$, the existence and uniqueness of the solution of the SDEs $f : Q \times X^{(\cdot)} \to \mathbb{R}^{d(\cdot)}$, $\sigma : Q \times X^{(\cdot)} \to \mathbb{R}^{d(\cdot) \times m(\cdot)}$ are bounded and Lipschitz continuous in z.
  Assumption about non-Zeno executions: if we denote $N_t(\omega) = \sum I_{(t \geq T_k)}$ then for every starting point $x \in X$, $\mathbf{E}^x N_t < \infty$, for all $t \in \mathbb{R}_+$.
  Assumption about the transition measure and the transition rate function:

  (A) $\lambda : X \to \mathbb{R}_+$ is a measurable function such that $t \to \lambda(x_t^i(\omega_i))$ is integrable on $[0, \varepsilon(x^i))$, for some $\varepsilon(x^i) > 0$, for each $z^i \in X^i$ and each ω_i starting at z^i.
  (B) for all A Borel measurable set, $R(\cdot, A)$ is measurable; for all $x \in \overline{X}$ the function $R(x, \cdot)$ is a probability measure; $R(x, \{x\}) = 0$ for $x \in X$.

The communication structure from (iv) is given by a structured set *labels*, each label $l \in L$ is a set $l = \{l^s, l^r\}$. There is a function which assigns a label to each jump from J, but we do not use it here.

Theorem 1. *Every aCPS is a strong Markov process.*

We assume that M is *transient* [13]. The transience of M means that any process trajectory which will visit a Borel set of the state space it will leave it after a finite time.

In this model, the environment is represented in two forms: the noise perturbation of system behavior in each location, and the information provided by the coloring functions. One can suppose the following scenario: The multi-sensorial perception of the changes in the environment is input to a safety analysis process. The results of the safety analysis consist of probabilities in reaching dangerous state sets. The safety evaluations of these probabilities are communicated to the human operator in the form of colored state sets. A yellow colored region means that a change (discrete transition) is recommended. A red colored region means that the automatic control must act.

A *multi-agent stochastic cyber-physical system* is a finite set of aPCS which can communicate pairwise using a common set of communication channels.

3   SafAL, the Safety Analysis Logic

In this section, we define a model theoretic logic, the safety analysis logic (SafAL), for specifying probabilistic safety properties of the aCPS model defined in the previous section. This is a qualitative approach that can complement the already existing numerical approaches. The qualitative reasoning provides a global and symbolic expression of the reach set probabilities, which is a good starting point for numerical evaluations. This logic is necessary in the formal specification of the coloring maps, as well as for the specification of normative prescriptions to the human operator.

3.1   Syntax and Functional Semantics

We depart from a variant of Larsen and Skou's probabilistic modal logic [20], a logic that has also inspired the real valued logic for the (discrete) labeled Markov Processes from [15]. Our approach differs fundamentally. The formulas of the logic are upper bounds for probabilities of reachable sets. In [15], the meaning of a formula is some measurable function.

The syntax is constructed from a formal description of a Markov process. That means we have a logic language where we can specify concepts like probability space, random variables, transition probabilities.

The main design scheme is based on the following principles. The system is modeled by a general Markov process. The sets of states are coded by their indicator functions. Obviously, these are elements of $\mathbf{B}(X)$. The application of the kernel operator on these functions generates the probabilities of the events that the system trajectories hit the respective sets.

The vocabulary of the logic is given by a family of measurable sets in a Lusin space. Each set is represented by its indicator function. For example, the interval $A = [0, 1/2]$ is represented, in the logic, by the function 1_A, which in each point x takes the value 1 if $0 \leq x \leq 1/2$ and the value 0 otherwise. The union of two disjoint sets A and B will be represented by the function $1_A + 1_B$. The intersection of two sets A and B will be represented by the function $\inf(1_A, 1_B)$. The complementary of the set A is represented by $1 - 1_A$.

We consider a linear space of bounded measurable functions, ranged over by the variable f. We define the terms by the following rules:

- the atomic terms are given by 1 or $\mho.f$, where $\mho$ is an *action operator*
- any other term is obtained from the atomic terms using:

$$g := g + g' \,|\, g - g' \,|\, \inf(g, g') \,|\, \sup_{n \in \mathbb{N}} g_n$$

The set of terms is denoted by $\mathcal{T}$.

A *reach type* formula is a statement of the form $g \leq v$, where g is a term and v is a real in $[0, 1]$. Other reach type formulae are obtained using the usual Boolean operators.

A *color* consists of a sub-interval $[a, b]$ of $[0, 1]$ and a subset C of the vocabulary such that $\mho.f$ takes values in $[a, b]$, for all $f \in C$.

A *safety formula* is obtained using the predefined predicate ∂ and the logical operator ∇ (which can not be nested).

The meaning of ∂ is that a discrete transition is triggered. A formula of the form ∇P means that it is "obligatory" [2] that the predicate P will be fulfilled. In particular, $\nabla \partial$ means that a jump is requested. The formula $\neg \nabla \neg P$ means that the predicate P is "permitted", $\nabla \neg P$ means that the predicate P is "forbidden".

The semantic domain consists of a Markov process $M = (x_t, P_x)$ satisfying the hypotheses of Section 2 and a countable set of Boolean variables *Prop*.

We consider only those bounded measurable functions that are indicator functions of measurable sets of states. For simplicity, consider a term, which contains only a bounded measurable function f. Intuitively, a term denotes a function that, when applied to a state x, provides the probability of all trajectories, starting from x, reaching a set 'indicated' by f. The reachability property is formulated relatively to several sets of states, then the term is formed with their indicator functions.

The interpretation of a term $f \in \mathcal{T}$ is a function $f : X \rightarrow \mathbb{R}$, which is a measurable bounded function. The interpretation $\Im$ of the atomic terms is given by:

$$\Im(1) = 1, \quad \Im(\mho.f) = \mho.f$$

where, for all $x \in X : 1(x) = 1$, $(\mho.f)(x) = \int [\int_0^\infty f(x_t(\omega)) dt] P_x(d\omega)$
The infimum and supremum are defined pointwise.
The following characterization of the action of $\mho$ to a term g is insightful

$$(\mho.f)(\cdot) = E.[\int_0^\infty f(x_t) dt] = \int_0^\infty P_t f(\cdot) dt = V f(\cdot). \tag{1}$$

The terms are statistical statements about sets in the state space. An atomic term is the expectation of the random variable provided by the "visits" of a target set.

The formal semantics of the ∇ operator is defined considering a family of Markov processes $\mathcal{P}$ and a "normative" relation $\eqsim \subseteq \mathcal{P} \times \mathcal{P}$ and a valuation function $V : Prop \times \mathcal{P} \to \{0, 1\}$. The $P \eqsim Q$ denotes that P is an alternative norm to Q. The function V assign a truth value to any variable in a MAPS $\mathcal{P}$ and we write $M \ P \vDash A$ whenever $V(A, P) = 1$.

$M \ P \vDash A$ iff $V(A, P) = 1$.
$M \ P \vDash \neg A$ iff not $M \ P \vDash A$.
$M \ P \vDash \nabla A$ iff $.\forall Q \in \mathcal{P}$ (if $P \eqsim Q$ then $M \ Q \vDash A$).

Example 1. Consider the case of an aircraft for which we want to check that the probability to reach the sphere $S(u, 2)$ starting from an initial point x is less than 0.01. We can consider a Markov process in the Euclidean space modelling the aircraft dynamics. The probability is given by the following SafAL formula $\mho.(1_{S(u,2)})(x) \leq 0.01$ which in the above semantics means $E_x[\int_0^\infty 1_{S(u,2)}(x_t)dt] \leq 0.01$. This formula appears frequently in the mathematical models used in air traffic control.

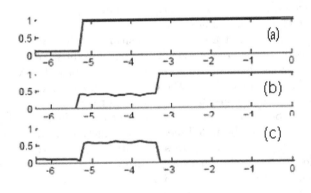

Fig. 4.

Example 2. The SafAL formula $\mho.1_A \geq 0.25 \supset \nabla \eqsim$ describes a yellow colored state set, where the dangerous area is described by the term A and the danger level for a reach probability is 0.25.

Example 3. The graphs from Fig. 4 depict SafAL terms (i.e. functions) that can be used for defining colors. Each graph depict the values of the probability of a discrete transition relative to the time of the first discrete transition. In the case (a) it is depicted the probability of a required discrete transition. In the case (b) it is shown the user ability to trigger a controlled transition, in the form of a probability. The case (c) illustrates the probability distribution for a supervisory controller to react, and it depends on the previous probabilities.

3.2   A Formal Semantics Based on Symmetries

Symmetry reduction is an efficient method for exploiting the occurrence of repli-
cation in discrete probabilistic model checking [19]. The verification of a model
can be then performed for a bisimilar quotient model, which is up to factorial
smaller. This is why, in this section, we explore the possibility of an alternative
semantics of SafAL based on symmetries.

Let $\mathcal{S}(X)$ be the group of all homeomorphisms $\varphi : X \to X$, i.e. all bijective
maps φ such that φ, φ^{-1} are $\mathcal{B}(X)$-measurable. When X is finite, $\mathcal{S}(X)$ is the
set of (finite) permutations of X.

Any permutation[2] of X induces a permutation of the group of measurable
functions (in particular of the terms) as follows. Let $* : \mathcal{S}(X) \to Perm[\mathbf{B}(X)]$
be the *action* $\mathcal{S}(X)$ to $\mathbf{B}(X)$ defined by $*(\varphi) = \varphi^* : \mathbf{B}(X) \to \mathbf{B}(X)$ where φ^* is
the linear operator on $\mathbf{B}(X)$ given by

$$\varphi^* f = f \circ \varphi. \tag{2a}$$

The range of $*$ is included in $Perm[\mathbf{B}(X)]$ (the permutation group of $\mathbf{B}(X)$).
This fact is justified by the invertibility of φ^*. The invertibility of φ^* can be
derived from the bijectivity of $\varphi \in \mathcal{S}(X)$ because it is clear that $(\varphi^*)^{-1} = (\varphi^{-1})^*$.
Then φ^* can be thought of as a symmetry of $\mathbf{B}(X)$ for each φ given in the
appropriate set (see also the appendix).

Consider now a Markov process M, as in the Section 2 and the excessive
function cone $\mathcal{E}_M$ (clearly a semigroup included in $\mathbf{B}(X)$). We can not define
the action of $\mathcal{S}(X)$ to $\mathcal{E}_M$ using formula (2a) because the result of composition
in (2a) is not always an excessive function.

Therefore it is necessary to consider some subgroups of permutations of the state
space such that we can define the action of these subgroups on the semigroup of
the excessive functions $\mathcal{E}_M$.

We consider the *maximal subgroup of permutations* of the state space X, de-
noted by $\mathcal{H}$, such that we can define the *action of* $\mathcal{H}$ *to* $\mathcal{E}_M$ $* : \mathcal{H} \to Perm[\mathcal{E}_M]$
defined as the appropriate restriction of (2a). The elements of $\mathcal{H}$ 'preserve'
through '$*$' the excessive functions, or, in other words, the stochastic specifications of the system.

In the spirit of [19], the elements of $\mathcal{H}$ are called *automorphisms*. Note that in
[19], the automorphisms are permutations of the state space, which preserve the
transition system relation. For the Markov chains, the automorphisms defined in
[19] preserve the probability transition function. For the case of continuous-time
continuous space Markov processes, a transition system structure is no longer
available (the concept of next state is available only for Markov chains). Therefore, it should be the case that the definition of the concept of automorphism to
be different: An automorphism must preserve the probabilistic dynamics of the
system. To express formally this idea, we need to use global parameterizations
of Markov processes different from transition probabilities, which are local and

---

[2] Here, permutation is used with the sense of one-to-one correspondence or bijection.

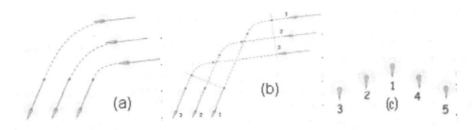

Fig. 5.

depend on time. This is the reason why we have defined these automorphisms as maps which preserve the excessive functions.

Using $\mathcal{H}$, an equivalence relation $\mathcal{O} \subset X \times X$, called *orbit relation*, can be defined on the state space X as follows.

Definition 1. *Two states x, y are in the same orbit, written $x \mathcal{O} y$, if and only if there exists an automorphism $\varphi \in \mathcal{H}$ such that $\varphi(x) = y$.*

Let us denote by $[x]$ the equivalence class containing the point x in X. The equivalent classes of $\mathcal{O}$ are called *orbits*. It is clear that an orbit $[x]$ can be described as $[x] = \{\varphi(x) | \varphi \in \mathcal{H}\}$. Let $X/\mathcal{O}$ denote the set of orbits, and let $\Pi_{\mathcal{O}}$ the canonical projection

$$\Pi_{\mathcal{O}} : X \to X/\mathcal{O}, \; \Pi_{\mathcal{O}}(x) = [x]. \tag{3}$$

The space $X/\mathcal{O}$ will be equipped with the quotient topology by declaring a set $A \subset X/\mathcal{O}$ to be open if and only if $\Pi_{\mathcal{O}}^{-1}(A)$ is open in X. It is clear now that $\Pi_{\mathcal{O}}$ is a continuous map with respect to the initial topology of X and the quotient topology of $X/\mathcal{O}$.

Example 4. In Figure 5 there are illustrated some basic situations for using symmetries and permutations. In the (a), the agent trajectories on the side are symmetric in respect with the middle agent trajectory. In the case (b), the symmetric agents from case (a) are permuted. In the case (c), the agent trajectories denoted by 3, 4 and 5 can be obtained by symmetry from the configuration formed with the agents 1 and 2: agent 3 is the symmetric of agent 1 with respect with agent 2; then, considering agent 1 trajectory as reference, agent 4 is the symmetric of agent 2 and agent 5 is the symmetric of agent 3.

Example 5. Consider that in Figure 5.(a) there are depicted three cars on a motorway. The symmetry approach allows one to consider only two cars placed in lanes next to each other.

Consider an automorphism $\varphi \in \mathcal{H}$.

Definition 2. *A term g is called φ-symmetric in $x, y \in X$ if*

$$\varphi(x) = y \Rightarrow g(x) = g(y). \tag{4}$$

The φ-symmetry property of a term gives rise to a new concept of *satisfaction* for a formula.

Definition 3. *A formula $g \leq v$ is equally satisfied in $x, y \in X$ if there exists an automorphism $\varphi \in \mathcal{H}$ such that g is φ-symmetric in $x, y \in X$.*

4   Model Checking Safety Properties

In this section, we investigate the issue of model checking the SafAL formulae. Because SafAL is a qualitative approach to probabilistic risk analysis, the model checking should be defined in an abstract way, rather than computationally. We establish mathematical properties relative to bisimulation and reachability analysis. In the first step, it is natural to consider model checking relative to any target set regardless their coloring.

4.1   Safety Equivalence

The computational equivalence of processes (or bisimulation) is the traditional tool for reducing the complexity of the system state space. In the following, we define this bisimulation for a class of strong Markov processes in an analytical setting. Safety properties can be much more easily checked using a bisimilar system abstraction as illustrated in [11]. In our approach, computational equivalence means equal risk, and two states (safety) are bisimilar if they carry equal risk. We develop a series of mathematical results that constitute the key for the risk assessment. Roughly speaking, these results allow to interpret the risk in terms of the mathematical potentials associated to a Markov process.

For a continuous time continuous space Markov process M with the state space X, an equivalence relation $\mathcal{R}$ on X is a *(strong) bisimulation* if for $x\mathcal{R}y$ we have

$$p_t(x, A) = p_t(y, A), \forall t > 0, \forall A \in \mathcal{B}(X/_R) \tag{5}$$

where $p_t(x, A)$, $x \in X$ are the transition probabilities of M and $\mathcal{B}(X/_R)$ represents the σ-algebra of measurable sets closed with respect to $\mathcal{R}$. This variant of strong bisimulation considers two states to be equivalent if their 'cumulative' probability to 'jump' to any set of equivalent classes that this relation induces is the same. The relation (5) is hard to be checked in practice since the time t runs continuously. Therefore, to construct a robust bisimulation relation on M it is necessary to use other characterizing parameters of M, such that formula (5) becomes a particular case of this new bisimulation.

In the following we briefly present the concept of bisimulation defined in [11]. This concept is more robust because it can be characterized by an interesting pseudometric [11].

Let $E \in \mathcal{B}(X_\Delta)$ be a measurable set. Let us consider $T_E = \inf\{t > 0 | x_t \in E)$, the first time at which a given process "hits" a given subset E of the state space.

It is possible to define a linear operator on $\mathbf{B}(X)$ (set of measurable bounded functions), denoted P_E by

$$P_E f(x) = P_x[f(x_{T_E})|T_E < \infty]. \tag{6}$$

If f is excessive, then so is $P_E f$. In particular, $P_E 1(x) = P_x[T_E < \infty]$ is excessive for any $E \in \mathcal{B}(X_\Delta)$. It can be shown that this function represents the probability measure of the set of process trajectories which hit the target set E, in infinite horizon time [11].

Suppose we have given a Markov process M on the state space X, with respect to a probability space $(\Omega, \mathcal{F}, \mathbf{P})$. Assume that $\mathcal{R} \subset X \times X$ is an equivalence relation such that the quotient process $M|_{\mathcal{R}}$ is still a Markov process with the state space $X/_{\mathcal{R}}$, with respect to a probability space $(\Omega, \mathcal{F}, \mathbf{Q})$. That means that the projection map associated to $\mathcal{R}$ is a Markov function.

Definition 4. *A relation $\mathcal{R}$ is called a behavioral bisimulation on X if for any $A \in \mathcal{B}(X/_{\mathcal{R}})$ we have that*

$$\mathbf{P}[T_E < \infty] = \mathbf{Q}[T_A < \infty]$$

where $E = \Pi_{\mathcal{R}}^{-1}(A)$ (i.e. the reach set probabilities of the process M and $M|_{\mathcal{R}}$ are equal).

Our first major assumption is that $X/_{\mathcal{O}}$ is a Lusin space. Often, this assumption can be checked, but there are some cases when $X/_{\mathcal{O}}$ fails to be Hausdorff (i.e. it is possible that two different orbits to share the same vicinity system). In these cases some minor modifications of X (changing, for example, the original topology) lead to a Hausdorff quotient space.

The main result of this section is that the orbit relation $\mathcal{O}$ is indeed a bisimulation relation defined on the state space X.

Theorem 2. *The orbit relation $\mathcal{O}$ is a behavioral bisimulation (as in the Definition 4).*

To prove this theorem we need some auxiliary results, which will be developed in the following.

Lemma 1. *If $f \in \mathcal{E}_M$ and $\varphi \in \mathcal{H}$ then*

$$P_E f = \varphi^*[P_F(\vartheta)] \tag{7}$$

where $F = \varphi(E)$; $\vartheta = \varphi^{-1*}f$ the action of '$*$' is given by (2a) and P_F is the hitting operator associated to F.

Proof of Lemma 1. It is known (Hunt's balayage theorem [3]) that

$$
\begin{aligned}
P_E f(x) &= \inf\{h(x)|h \in \mathcal{E}_M, h \geq f \text{ on } E\} \stackrel{\text{(if } \varphi \in \mathcal{H})}{=} \\
&= \inf\{h \circ \varphi^{-1}(\varphi(x))|h \in \mathcal{E}_M, h \circ \varphi^{-1} \geq f \circ \varphi^{-1} \text{ on } \varphi(E)\} \\
&= \inf\{k(\varphi(x))|k \in \mathcal{E}_M, k \geq f \circ \varphi^{-1} \text{ on } \varphi(E)\} \\
&= P_{\varphi(E)}(f \circ \varphi^{-1})(\varphi(x)) \\
&= P_{\varphi(x)}[(f \circ \varphi^{-1})(x_{T_{\varphi(E)}})|T_{\varphi(E)} < \infty].
\end{aligned}
$$

Remark 1. The equality (7) remains true for functions of the form $f_1 - f_2$ where f_1 and f_2 are excessive functions, and from there to arbitrary Borel measurable functions.

Proposition 1. *Let $g : X/_{\mathcal{O}} \to \mathbb{R}$ be a $\mathcal{B}(X/_{\mathcal{O}})$-measurable and let $E = \Pi_{\mathcal{O}}^{-1}(A)$ for some $A \in \mathcal{B}(X/_{\mathcal{O}})$. Then*

$$P_E f = \varphi^*[P_A f], \forall \phi \in \mathcal{H} \tag{8}$$

where $f : X \to \mathbb{R}$, $f = g \circ \Pi_{\mathcal{O}}$.

Proof of Prop. 1. If in Lemma 1, we let $f = g \circ \Pi_{\mathcal{O}}$, then $\vartheta = \varphi^{*-1} f = f \circ \varphi^{-1} = g \circ \Pi_{\mathcal{O}} \circ \varphi^{-1} = f$. More, $\varphi(\Pi_{\mathcal{O}}^{-1}(A)) = \Pi_{\mathcal{O}}^{-1}(A)$, so the proposition follows from the above lemma.

Formula (8) shows that the function $P_E f$ (where $f = g \circ \Pi_{\mathcal{O}}$) is constant on the equivalent classes with respect to $\mathcal{O}$. Then it makes sense to define a collection of operators (Q_A) on $(X/_{\mathcal{O}}, \mathcal{B}(X/_{\mathcal{O}}))$ by setting $Q_A g([x]) = P_E(g \circ \Pi_{\mathcal{O}})(x)$ where $E = \Pi_{\mathcal{O}}^{-1}(A)$. Proposition 1 allows to use any representative x of $[x]$ in the right side. It easy to check that $Q_A Q_B = Q_B$ if A and B are open sets of $X/_{\mathcal{O}}$ with $B \subset A$. Under some supplementary hypotheses one can construct a Markov process $M/_{\mathcal{O}} = (\Omega, \mathcal{F}, \mathcal{F}_t, [x]_t, Q_{[x]})$ with these hitting operators [13].

Now, we have all the auxiliary results needed to prove the Theorem 2.
Proof of the Th.2. If $E = \Pi_{\mathcal{O}}^{-1}(A)$ for some $A \in \mathcal{B}(X/_{\mathcal{O}})$ and we let $g \equiv 1$ then, for all $x \in X$

$$P_x[T_E < \infty] = Q_{[x]}[T_A < \infty]. \tag{9}$$

Formula (9) illustrates the equality of the reach set probabilities, i.e. $\mathcal{O}$ is a bisimulation relation.

4.2   Logical Characterization of Safety Bisimulation

Theorem 3 (Full Abstraction Theorem). *Any two states $x, y \in X$ are bisimilar (through $\mathcal{O}$) if and only if, for any SafAL formula is equally satisfied by x and y.*

Proof of the Th. 3.
Necessity:
$x \mathcal{O} y$ implies that there exists $\overline{\varphi} \in \mathcal{H}$ such that $\overline{\varphi}(x) = y$. Since $\overline{\varphi} \in \mathcal{H}$, for all $g \in \mathbf{B}(X)$ we have from Lemma 1 (taking $E = \{\Delta\}$ and $\overline{\varphi}\{\Delta\} = \Delta$) that $Vg(x) = V(g \circ \overline{\varphi}^{-1})(\overline{\varphi}(x))$. Or, taking $\varphi = \overline{\varphi}^{-1}$

$$Vg(x) = (Vg \circ \varphi)(x) \tag{10}$$

and using the fact any excessive function f is the limit of an increasing sequence of potentials (by Hunt theorem [3]) we can make the following reasoning. For a stochastic specification $f \in \mathcal{E}_M$ there exists a sequence $(g_n) \subset \mathbf{B}(X)$ such that Vg_n is increasingly converging to f. Then, from (10), we obtain that $(f \circ \varphi)(x) = \uparrow \lim Vg_n(\varphi(x)) = \uparrow \lim Vg_n(x), \forall x \in X$, i.e., we get that $f(x) = (f \circ \varphi)(x)$,

$\forall f \in \mathcal{E}_M$. Therefore, the evaluations of each stochastic specification f in x and y are equal when $x\mathcal{O}y$. Then the result is true also for $f \in \mathcal{T}$, since any measurable function can be represented as the difference of two excessive function.

Sufficiency:
In this case, we have to show that if for each $f \in \mathcal{T}$ there exists $\varphi \in \mathcal{H}$ such that (4) is true, then $x\mathcal{O}y$. This statement is straightforward.

The Full Abstraction Theorem establishes that our model is correct and complete for the safety analysis logic. It provides new insights to the bisimulation relation $\mathcal{O}$, as follows.

Two states are equivalent when, for all system trajectories passing them some relevant probabilistic properties are evaluated to be the same. This computational copy of a state is given by the permutation φ from (4).

Corollary 1. The action of $\mathcal{H}$ to $\mathcal{E}_M$ can be restricted as the action of $\mathcal{H}$ to $\mathcal{P}_M$, i.e. $* : \mathcal{H} \times \mathcal{P}_M \to \mathcal{P}_M$ given by (2a).

This corollary is a direct consequence of the fact that $\mathcal{P}_M$ generates the cone $\mathcal{E}_M$. Then in the definition of $\mathcal{O}$, we can work not with excessive functions, but with potentials. This means that we can give the following characterization of the orbit relation.

Proposition 2. $x\mathcal{O}y$ if and only if they have the 'same potential', i.e. there exists $\varphi \in \mathcal{H}$ such that $Vf(x) = Vf(y)$, $y = \varphi(x)$ for all $f \in \mathbf{B}(X)$

Corollary 2. If $x\mathcal{O}y$ then there exists $\varphi \in \mathcal{H}$ such that $\varphi(x) = y$ and $p_t(x, \varphi^{-1}(A)) = p_t(y, A), \forall t \geq 0, \forall A \in \mathcal{B}$.

Summarizing, the model checking problem provides a very good motivation for the colored model. Without using colors, the only safety bisimilar states exist only for the systems that exhibit symmetries. Intuitively, two states are bisimilar only if they are spatially symmetric. Using colors, the bisimulation concept is coarser because the risk is considered only for colored sets. For example, two states can be colorly bisimilar even they are not spatially symmetric. For the car example, a state situated in the vicinity of a colored set, but characterized by a small velocity can be safety equivalent with a state situated far from the colored set but characterized by high velocity. Moreover, because the coloring functions depend on time, the safety equivalence for the colored model varies over the time. In other words, two safety bisimilar states can not be bisimilar anymore a few seconds later.

5   Final Remarks

Conclusions
In this paper, we have proposed the multi-dimensional system co-engineering framework, consisting of a stochastic multi-agent model, a formal logic for expressing safety properties and a foundational study of the basic formal verification concepts of bisimulation and model checking. An agent is modeled as an

evolution of a cyber-physical system, which, in turn is an extension of a stochastic hybrid system. Agent models have been developed for discrete probabilistic systems [16] and hybrid systems [1]. Moreover, model checking methodologies have been developed for these systems. However, these methods can not be extended for agent models of systems which are simultaneously hybrid and stochastic. In this case, essential systems properties are lost and new properties emerge, as described within Hilbertean formal methods [4,5].

Examples of lost properties include:

- The uniqueness of a continuous trajectories that start from a given point;
- The availability of the "next state" concept
- The representation of the transition probabilities in the compact form of a matrix. Instead, this situation leads to the use of linear operators. Consequently, their specification logics should be based on a different semantics.

The following situations can be considered as emergent properties:

- In the description of the system behavior, one is constrained to use only measurable sets of states and measurable sets of trajectories. Therefore, a specification logic based on such principles needs to be introduced.
- The reachability properties can not be expressed, as in the discrete case, using only the transition probabilities. Instead, we have to consider measurable sets of trajectories that visit a target set of states. This situation conducts to possibly unpleasant consequences: the model checking techniques developed for deterministic hybrid systems or for discrete probabilistic systems are not usable anymore.

To the authors knowledge the problems presented in this paper and the proposed solutions are new.

The existence of a fully abstract model, but still very general and constructive, forms the basis for future automated reasoning systems.

Related work

A model of agents as deterministic hybrid systems that communicate via shared variables is implemented in the Charon system (see [1] and the references therein). For discrete time and probabilistic agents, there exist well developed models [17]. However it is very difficult to model the agents as stochastic hybrid systems, especially because of the emergent properties presented before.

Symmetries have been used by Frazzoli and coauthors (see [18] and the references therein) in the optimal control of single agent, deterministic hybrid systems.

Bisimulation has got a large palette of definitions for discrete systems, and, similarly, there exist different definitions in the continuous and hybrid case. A categorical definition is proposed in [12], and non-categorical variants are introduced and investigated in [7], [11]. Other approaches to formal verification of probabilistic systems, like labelled Markov processes [14], consider automata models which are not agent oriented. The full abstraction theorem from section 6 extend a similar result established for discrete probabilistic automata.

Note that, in contrast with the action operator defined in the probabilistic modal logic for labeled Markov processes [14], the SafAL operator is defined using the time.

The SafAL can be fruitfully applied to performance analysis. In [10], it is shown that the expressions (1), i.e. the semantics of some SafAL formulas, represent performance measures for the fluid models of communication networks. Moreover, in the cited paper, it is developed a model checking strategy for a set of formulae that belong to SafAL against strong Markov processes, which enrich the formal verification toolset of MScE.

Future work
In a following paper we will extended SafAL to include inter-agent communication and develop an operational semantics for it. Considering the efficient model checking methods based on symmetry reduction [19], it is natural to further investigate developing similar numerical methods for SafAL. Application domains where MScE can be used include aerospace engineering[3], air traffic control and automotive industry.

More background material on stochastic processes and stochastic hybrid systems can be found in an early version [9] of this paper which is available on-line[4].

Acknowledgments. This work was partly funded by the EPSRC project EP/E050441/1. We thank the anonymous reviewers for useful comments.

References

1. Alur, R., Grosu, R., Hur, Y., Kumar, V., Lee, I.: Modular specification of hybrid systems in CHARON. In: Lynch, N.A., Krogh, B.H. (eds.) HSCC 2000, vol. 1790, pp. 6–19. Springer, Heidelberg (2000)
2. Barringer, H.: The Future of Imperative Logic. British Colloquium for Theoretical Computer Science (1990)
3. Boboc, N., Bucur, G., Cornea, A.: Order and Convexity in Potential Theory. H-Cones. Lecture Notes in Math., vol. 853. Springer, Heidelberg (1992)
4. Bujorianu, M.C., Bujorianu, M.L.: Towards Hilbertian Formal Methods. In: Proc. of Conf. on Application of Concurrency to System Design, IEEE Press, Los Alamitos (2007)
5. Bujorianu, M.C., Bujorianu, M.L.: An Integrated Specification Framework for Embedded Systems. In: Proc. of SEFM. IEEE Press, Los Alamitos (2007)
6. Bujorianu, M.C., Bujorianu, M.L.: Towards a Formal Framework for Multidimensional Codesign. Technical Report TR-CTIT-08-21 Centre for Telematics and Information Technology, University of Twente (2008), http://eprints.eemcs.utwente.nl/12108/
7. Bujorianu, L.M., Bujorianu, M.C.: Bisimulation, Logic and Mobility for Markovian Systems. In: Proc. of Eighteenth International symposium on Mathematical Theory of Networks and Systems (2008)

---

[3] *Engineering Autonomous Space Software* Project, EPSRC EP/F037201/1, http://gow.epsrc.ac.uk/ViewGrant.ASPx?Grant=EP/F037201/1
[4] http://eprints.eemcs.utwente.nl/12112/

8. Bujorianu, M.L.: Extended Stochastic Hybrid Systems and their Reachability Problem. In: Alur, R., Pappas, G.J. (eds.) HSCC 2004. LNCS, vol. 2993, pp. 234–249. Springer, Heidelberg (2004)
9. Bujorianu, L.M., Bujorianu, M.C.: Bisimulation, Logic and Reachability Analysis for Markovian Systems. Technical Report TR-CTIT-08-23 Centre for Telematics and Information Technology, University of Twente (2008)
10. Bujorianu, M.L., Bujorianu, M.C.: Model Checking for a Class of Performance Properties of Fluid Stochastic Models. In: Horváth, A., Telek, M., et al. (eds.) EPEW 2006. LNCS, vol. 4054, pp. 93–107. Springer, Heidelberg (2006)
11. Bujorianu, M.L., Lygeros, J., Bujorianu, M.C.: Abstractions of Stochastic Hybrid System. In: Proc. 44th Conference in Decision and Control. IEEE Press, Los Alamitos (2005)
12. Bujorianu, M.L., Lygeros, J., Bujorianu, M.C.: Bisimulation for General Stochastic Hybrid Systems. In: [21], pp. 198–216
13. Blumenthal, R.M., Getoor, R.K.: Markov Processes and Potential Theory. Academic Press, London (1968)
14. Desharnais, J., Edalat, A., Panangaden, P.: A Logical Characterization of Bisimulation for Labeled Markov Processes. In: LICS, pp. 478–487 (1998)
15. Desharnais, J., Gupta, V., Jagadeesan, R., Panangaden, P.: Metrics for Labelled Markov Systems. In: Baeten, J.C.M., Mauw, S. (eds.) CONCUR 1999. LNCS, vol. 1664, pp. 258–273. Springer, Heidelberg (1999)
16. Fisher, M., Bordini, R.H., Hirsch, B., Torroni, P.: Computational Logics and Agents: A Roadmap of Current Technologies and Future Trends. Computational Intelligence 23(1), 61–91 (2007)
17. Fisher, M., Ballarini, P., Wooldridge, M.: Uncertain Agent Verification through Probabilistic Model-Checking. In: Proceedings of 3rd International Workshop on Safety and Security in Multiagent Systems (SASEMAS 2006) (2006)
18. Frazzoli, E., Bullo, F.: On Quantization and Optimal Control of Dynamical Systems with Symmetries. Proc. IEEE Conf. on Decision and Control 1, 817–823 (2002)
19. Kwiatkowska, M., Norman, G., Parker, D.: Symmetry Reduction for Probabilistic Model Checking. In: Ball, T., Jones, R.B. (eds.) CAV 2006. LNCS, vol. 4144, pp. 234–248. Springer, Heidelberg (2006)
20. Larsen, K.G., Skou, A.: Bisimulation through Probabilistic Testing. Information and Computation 94, 1–28 (1991)
21. Morari, M., Thiele, L. (eds.): HSCC 2005. LNCS, vol. 3414. Springer, Heidelberg (2005)
22. McCall, J.C., Trivedi, M.M.: Driver Behavior and Situation Aware Brake Assistance for Intelligent Vehicles. Proceedings of the IEEE 95(2), 374–387 (2007)
23. Pola, G., Bujorianu, M.L., Lygeros, J., Di Benedetto, M.D.: Stochastic Hybrid Models: An Overview with applications to Air Traffic Management. In: Proccedings Analysis and Design of Hybrid Systems, IFAC ADHS 2003, pp. 45–50. Elsevier IFAC Publications (2003)
24. Wikipedia Cyber-physical system,
   http://en.wikipedia.org/wiki/Cyber-physicalsystems

Appendix: Background on Stochastic Processes

Let us consider $M = (x_t, P_x)$ a strong Markov process with the state space X. Let $\mathcal{F}$ and $\mathcal{F}_t$ be the appropriate completion of σ-algebras $\mathcal{F}^0 = \sigma\{x_t | t \geq 0\}$

and $\mathcal{F}_t^0 = \sigma\{x_s | s \leq t\}$. $\mathcal{F}_t$ describes the history of the process up to the time t. Technically, with any state $x \in X$ we can associate a natural probability space $(\Omega, \mathcal{F}, P_x)$ where P_x is such that its initial probability distribution is $P_x(x_0 = x) = 1$. The *strong Markov* property means that the Markov property is still true with respect to the stopping times of the process M. In particular, any Markov chain is a strong Markov process.

We adjoin an extra point Δ (the cemetery) to X as an isolated point, $X_\Delta = X \cup \{\Delta\}$. The existence of Δ is assumed in order to have a probabilistic interpretation of $P_x(x_t \in X) < 1$, i.e. at some 'termination time' $\zeta(\omega)$ when the process M escapes to and is trapped at Δ. As usual, $\mathbf{B}(X)$ denotes the set of bounded real functions on X.

Suppose that the following hypotheses are fulfilled.

1. M paths are right-continuous with left limits (the cadlag property).
2. X is equipped with Borel σ-algebra $\mathcal{B}(X)$ or shortly $\mathcal{B}$. Let $\mathcal{B}(X_\Delta)$ be the Borel σ-algebra of X_Δ.
3. The operator semigroup of M maps $\mathbf{B}(X)$ into itself.

- The set $\mathbf{B}(X)$ is the Banach space of bounded real measurable functions defined on X, with the sup-norm $||\varphi|| = \sup_{x \in X} |\varphi(x)|$, $\varphi \in \mathbf{B}(X)$.
- The semigroup of operators (P_t) is given by

$$P_t f(x) = E_x f(x_t) = \int f(y) p_t(x, dy), t \geq 0 \tag{11}$$

where E_x is the expectation with respect to P_x and $p_t(x, A)$, $x \in X$, $A \in \mathcal{B}$ represent the transition probabilities, i.e. $p_t(x, A) = P_x(x_t \in A)$. The semigroup property of (P_t) can be derived from the Chapman-Kolmogorov equations satisfied by the transition probabilities.

- A function f is *excessive* with respect to the semigroup (P_t) if it is measurable, non-negative and $P_t f \leq f$ for all $t \geq 0$ and $P_t f \nearrow f$ as $t \searrow 0$.

To the operator semigroup, one can associate the *kernel operator* as

$$Vf = \int_0^\infty P_t f dt, f \in \mathbf{B}(X) \tag{12}$$

The kernel operator is the inverse of the opposite of the infinitesimal operator associate to M.

Remark 2. The state space X can be chosen to be an analytic space (as the most general case), but we restrict ourself to the case of a Lusin space since our work is motivated by a multi-agent model where every agent is a realization of a stochastic hybrid systems who have, in most of the cases, Lusin state spaces.

Revisiting Satisfiability and Model-Checking for CTLK with Synchrony and Perfect Recall

Cătălin Dima

LACL, Université Paris Est – Université Paris 12,
61 av. du Général de Gaulle, 94010 Créteil, France

Abstract. We show that CTL with knowledge modalities but without common knowledge has an undecidable satisfiability problem in the synchronous perfect recall semantics. We also present an adaptation of the classical model-checking algorithm for CTL that handles knowledge operators.

1 Introduction

Combinations of temporal logics and epistemic logics offer a useful setting for the specification and analysis of multi-agent systems. They have been successfully utilized for model-checking protocols like the Alternating Bit Protocol [RL05], or the Chaum's Dining Cryptographers Protocol [vdMS04, KLN+06], whose functioning is related with participants' knowledge of the system state.

Epistemic temporal logics have been studied since the mid-eighties, starting with [HV86, HV89]. These two studies led to the identification of 96 different logics, distinguished by semantics and/or the presence of common knowledge operators, and concern a number of decidability and undecidability results for the satisfiability problem in those logics. In particular, it is shown that Linear Temporal Logic (LTL) with knowledge modalities and no common knowledge has a decidable satisfiability problem in a synchronous and perfect recall semantics. However, the results proved in [HV89, HV86] only concern extensions of LTL, and though both studies mention also some results on branching logics, neither of the two concentrates on proving (un-)decidability for the epistemic extensions of branching time logics.

In this paper, we study the Computational Tree Logic with knowledge operators *and without common knowledge*, a logic that we denote as CTLK. We show that, contrary to the result on LTL, satisfiability within CTLK is undecidable under the synchronous and perfect recall semantics. This result contradicts the claim in [HV86] following which this logic (denoted there KB_n) would be decidable in nonlinear time.

Our proof of the undecidability of CTLK satisfiability is somewhat classical, in the sense that we code the computation of a Turing machine vertically in a tree. This proof technique was utilized many times in the literature for proving undecidability of various epistemic temporal logics, starting from [HV86] where it is proved that LTL *with common knowledge operators* and with various semantics

M. Fisher, F. Sadri, and M. Thielscher (Eds.): CLIMA IX, LNAI 5405, pp. 117–131, 2009.

has an undecidable satisfiability problem. We also cite the undecidability result of [vdM98] for LTL *with common knowledge*, which utilizes the same type of argument. Another paper which utilizes this argument is [vBP06], in which it is shown that several variants of branching-time logics *with common knowledge operators*[1] have an undecidable satisfiability problem. But, to our knowledge, this is the first time an epistemic temporal logic *without a common knowledge operator* is shown to have an undecidable satisfiability problem.

We also investigate here the model-checking problem for CTL with knowledge. The model-checking problem for (a generalized form of) a branching-time logic with knowledge operators and without common knowledge has been studied in [SG02], where the approach is to code the model-checking problem as a satisfiability problem in Chain Logic [ER66, Tho92].

We take here a direct approach, by adapting the classical model-checking algorithm of [CES86]. The extra procedure that is needed is a state labeling with knowledge formulas. This involves a subset construction on the given model, since one needs to identify all histories which may be identically observed by agent i, when one wants to label states with formulas involving the K_i modality. Note also that our approach is similar to the model-checking algorithm for LTL with knowledge from [vdMS99], which also involves a subset construction, optimized for achieving better complexity.

Our approach does not improve the worst-case complexity of the algorithm, since each nesting of knowledge operators induces an exponential explosion, thus leading to a nonelementary complexity. But we believe that our approach could be more practical for formulas with low nesting of knowledge operators. In the approach of [SG02], the system is first translated into the Chain Logic, which needs then to be coded into Monadic Second Order logic [Tho92], and then an MSO-based tool like Mona [EKM98] has to be applied. In such an approach, since the system coding creates some formula with quantifier alternation, some unnecessary determinization steps for the resulting Büchi automata are then needed. Our approach avoids this, as each non-knowledge operator requires only state relabeling, and no state explosion.

It is interesting to note that CTLK is not the only logic extending CTL in which satisfiability is undecidable but model-checking is decidable. The logic TCTL, a dense-time extension of CTL, bears the same problem [DW99].

The model-checking problem for a branching-time logic with knowledge operators has also been addressed in [LR06]. Their approach is to have state-based observation, and not trace-based observation, and this induces a PSPACE complexity of the model-checking problem. Note however that, in general, state-based observation is not a sychronous and perfect recall semantics.

We have not investigated here the possibility to adapt these results to other semantics, but we believe that our arguments can be extended to handle non-synchronous and/or non-perfect recall semantics.

---

[1] As stated in [vBP06] on page 5, $\mathcal{L}_{ETL}$ "contains all the [...] temporal and knowledge operators", that is, *common knowledge* is included too. Hence Theorem 24 refers to this branching temporal logic *with common knowledge*.

The paper is organized as follows: the next section gives the syntax and semantics of $CTLK_{prs}$. We then present the undecidability result in the third section, and the model-checking algorithm in the fourth section. We end with a section of conclusions and comments.

2   Syntax and Semantics of $CTLK_{prs}$

We recall here the syntax and the semantics of CTL, the Computational Tree Logic, with knowledge modalities. Our semantics is a synchronous and perfect recall semantics which is based on observability of atomic proposition values, rather than on an "abstract" observability mapping on system states [FHV04]. We also give the semantics in a "tree-automata" flavor, as the models of a formula are presented as trees – which are unfoldings of transition systems.

We first fix some notations to be used through the paper. Given any set A, we denote by A^* the set of finite sequences over A. Hence, $\mathbb{N}^*$ denotes the set of *finite sequences of natural numbers*. The prefix order on A^* is denoted $\preceq$, hence $abcd \preceq abcde \preceq abcde$. For a partial function $f : A \rightarrow B$, its *support* is the set of elements of A on which f is defined, and is denoted $\mathsf{supp}(f)$. The first projection of a partial function $f : A \rightarrow B_1 \times \ldots \times B_n$ is denoted $f|_{B_1}$; similar notations are used for all the projections of f.

Given a set of symbols AP, which will denote in the sequel the set of *atomic propositions*, an *AP-labeled tree* is a partial function $t : \mathbb{N}^* \rightarrow 2^{AP}$ that bears some additional properties, that we detail in the following. First, note that an element x in the support of t denotes a node of the tree, while $t(x)$ denotes the label of that node.

To be a tree, a mapping $t : \mathbb{N}^* \rightarrow 2^{AP}$ has to satisfy the following properties:

1. The support of t is prefix-closed: for all $x \in \mathsf{supp}(t)$ and all $y \preceq x$, $y \in \mathsf{supp}(t)$.
2. Trees are "full": for all $x \in \mathsf{supp}(t)$, if $xi \in \mathsf{supp}(t)$ for some $i \in \mathbb{N}$, then for all $0 \le j \le i$, $xj \in \mathsf{supp}(t)$.
3. Trees are infinite: for all $x \in \mathsf{supp}(t)$ there exists $i \in \mathbb{N}$ s.t. $xi \in \mathsf{supp}(t)$.

For example, the subset of integers $\{\varepsilon, 1^*, 121^*, 131^*, 2, 21^*\}$ is the support of a tree, whereas $\{\varepsilon, 1, 2, 221^*\}$ is not, as it does not satisfy neither the fullness property (for node 2) nor the infinity property (for node 1). Here, ε denotes the empty sequence of integers, and 1^* denotes the set $1^* = \{1^n \mid n \ge 0\}$.

We say that t_1 is *similar with* a tree t_2 and denote this $t_1 \simeq t_2$ if, intuitively, t_2 is a rearrangement of t_1. Formally, $t_1 \simeq t_2$ if there exists a bijection $\varphi : \mathsf{supp}(t_1) \to \mathsf{supp}(t_2)$ for which

- $\varphi(\varepsilon) = \varepsilon$.
- For all $x \in \mathsf{supp}(t_1)$ and $i \in \mathbb{N}$ for which $xi \in \mathsf{supp}(t_1)$ there exists $j \in \mathbb{N}$ such that $\varphi(x)j \in \mathsf{supp}(t_2)$ and $\varphi(xi) = \varphi(x)j$.
- For all $x \in \mathsf{supp}(t_1)$, $t_1(x) = t_2\big(\varphi(x)\big)$

A *finite path* in a tree t is a sequence of elements in the support of t, $(x_i)_{0 \le i \le k}$ with x_{i+1} being the immediate successor of x_i ($0 \le i \le k-1$) w.r.t. the prefix order. An *infinite path* is an infinite sequence $(x_k)_{k\geq 0}$ with x_{k+1} being the immediate successor of x_k for all $k \geq 0$. A (finite or infinite) path is *initial* if it starts with ε, the tree root.

Next we define the observability relations for each agent. These relations are given by subsets $AP_i \subseteq AP$ of atomic propositions. The values of atoms in AP_i are supposed to be observable by agent i, and no other atoms are observable by i. Informally, an agent i does not distinguish whether the current state of the system is represented by the node x or by the node y in the tree if:

 - x and y lie on the same level of the tree and
 - The sequence of atomic propositions that agent i can observe *along the initial path that ends in x* is the same as the sequence of atomic propositions i can observe *along the initial path that ends in y*.

The first requirement makes this semantics *synchronous*, as it codes the fact that any agent knows the current absolute time, and the second requirement gives the *perfect recall* attribute of this semantics, as it encodes the fact that each agent records all the observations he has made on the system state, and updates his knowledge based on his recorded observations.

Formally, given an AP-labeled tree t, a subset $AP_i \subseteq AP$ and two positions x, y we denote $x \sim_{AP_i} y$ if

1. x and y are on the same level in the tree, i.e., $x, y \in \mathsf{supp}(t)$, $|x| = |y|$,
2. For any pair of nodes $x', y' \in \mathsf{supp}(t)$ with $x' \preceq x$, $y' \preceq y$ and $|x'| = |y'|$ we have that $t(x') \cap AP_i = t(y') \cap AP_i$.

Figure 1 gives an example of a (finite part of a) tree and some pairs of nodes which are or are not related by the two observability relations $\sim_{AP_1}$ and $\sim_{AP_2}$.

The logic we investigate here, which is the Computational Tree Logic with knowledge operators and with a synchronous and perfect recall semantics, denoted in the following as $CTLK_{prs}$, has the following syntax:

$$\phi ::= p \mid \phi \wedge \phi \mid \neg\phi \mid A\bigcirc\phi \mid \phi\,A\mathcal{U}\,\phi \mid \phi\,E\mathcal{U}\,\phi \mid K_i\phi$$

The semantics of $CTLK_{prs}$ is given in terms of tuples (t, x) where t is an AP-labeled tree, $x \in \mathsf{supp}(t)$ is a position in the tree, $AP_i \subseteq AP$ are some fixed subsets $(1 \leq i \leq n)$, and $\sim_{AP_i}$ are the above-defined observability relations:

$(t, x) \models p$ if $p \in t(x)$

$(t, x) \models \phi_1 \wedge \phi_2$ if $(t, x) \models \phi_j$ for both $j = 1, 2$

$(t, x) \models \neg\phi$ if $(t, x) \not\models \phi$

$(t, x) \models A\bigcirc\phi$ if for all $i \in \mathbb{N}$ with $xi \in \mathsf{supp}(t)$, $(t, xi) \models \phi$

$(t, x) \models \phi_1\,A\mathcal{U}\,\phi_2$ if for any infinite path $(x_k)_{k\geq 0}$ in t with $x_0 = x$
                           there exists $k_0 \geq 1$ with $(t, x_{k_0}) \models \phi_2$
                           and $(t, x_j) \models \phi_1$ for all $0 \leq j \leq k_0 - 1$

$(t, x) \models \phi_1\,E\mathcal{U}\,\phi_2$ if there exists a finite path $(x_j)_{1\leq j\leq k_0}$ in t with $x_0 = x$,
                           $(t, x_{k_0}) \models \phi_2$ and $(t, x_j) \models \phi_1$ for all $0 \leq j < k_0$

$(t, x) \models K_i\phi$ if for any $y \in \mathsf{supp}(t)$ with $x \sim_{AP_i} y$ we have $(t, y) \models \phi$

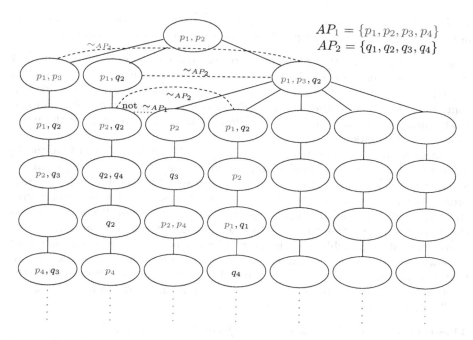

Fig. 1. An AP-tree, with some of the $\sim_{AP_1}$ relations and $\sim_{AP_2}$ relations represented as dashed lines. The dotted line shows two nodes that are not in $\sim_{AP_1}$ relation since the history that can be observed by agent 1 in both nodes is not the same, though the atomic propositions which are observed by 1 in each node is the same.

Remark 1. It is easy to note that for any pair of similar trees $t_1 \simeq t_2$, where the similarity relation is given by the bijection $\varphi : \mathsf{supp}(t_1) \to \mathsf{supp}(t_2)$, for any CTLK formula ϕ, and any position $x \in \mathsf{supp}(t_1)$, we have that

$$(t_1, x) \models \phi \text{ if and only if } (t_2, \varphi(x))$$

The usual abbreviations apply here too, in particular

$$E\Diamond\,\phi = true\ E\mathcal{U}\,\phi \qquad\qquad A\Box\,\phi = \neg\,E\Diamond\,\neg\phi$$
$$A\Diamond\,\phi = true\ A\mathcal{U}\,\phi \qquad\qquad E\Box\,\phi = \neg\,A\Diamond\,\neg\phi$$
$$P_i\phi = \neg K_i\neg\phi \qquad\qquad E\bigcirc\,\phi = \neg\,A\bigcirc\,\neg\phi$$

A formula ϕ is *satisfiable* if there exists a tree t and a position $x \in \mathsf{supp}(t)$ such that $(t, x) \models \phi$.

3   Undecidability of Satisfiability

This section presents our undecidability result for the satisfiability problem in $CTLK_{prs}$. The undecidable problem that will be simulated is the following:

Problem 1 (Infinite Visiting Problem). Given a deterministic Turing machine, once it starts with a blank tape, does it visit all the cells of its tape?

It is easy to see that this problem is co-r.e.: just construct a multi-tape TM simulator, in which one of the tapes serves for memorizing all the previously visited configurations. The simulator machine also fixes, at start, a marker at some cell, as a "guess" that the R/W head will never go beyond that marker. It also memorizes, for each reachable configuration, only the part of the input tape up to the marker.

The simulator machine simulates one step of the given TM, and first checks whether it has reached a final state, in which case it halts. If not, it checks whether the new configuration has ever been reached by checking whether it is present between the memorized configurations. If yes, then it goes into an error state in which it fills all the input tape with an error symbol. If not, it appends the current configuration to the memorized configurations and continues.

As it might be possible that the marker be reached during computation, that is, that the simulator reaches more cells during its computation that the amount that was initially guessed, then the simulator pushes the marker one cell to the right, and appends one blank cell to each of the memorized configurations.

The first result of this paper is the following:

Theorem 1. *Satisfiability of $CTLK_{prs}$ formulas is undecidable.*

Proof. The main idea is to simulate the (complement of the) Infinite Visiting Problem 1 by a $CTLK_{prs}$ formula ϕ_T. Similarly to [vdM98], a tree that would satisfy ϕ_T would have configurations of the given TM coded as inital paths, and the observability relations $\sim_{AP_1}$ and $\sim_{AP_2}$ would be used to code transitions between configurations.

So take a deterministic Turing machine $T = (Q, \Sigma, \delta, q_0, F)$ with $\delta : Q \times \Sigma \rightarrow Q \times \Sigma \times \{L, R\}$. Assume, without loss of generality, that the transitions in δ always change state, and that δ is total – hence T may only visit all its tape cells, or cycle through some configuration, or halt because trying to move the head to the left when it points on the first tape cell.

The formula ϕ_T will be constructed over the set of atomic propositions consisting of:

1. Four copies of Q, denoted Q, Q', $\overline{Q}$ and $\overline{Q}'$.
2. Two copies of Σ, denoted Σ and Σ'.
3. An extra symbol $\perp$.

The symbol $\perp$ will be used as a marker of the right end of the available space on the input tape on which T will be simulated, and its position will be "guessed" at the beginning of the simulation. The utility of the copies of Q and Σ is explained in the following, along with the way computations of T are simulated.

The computation steps of T are simulated by inital paths in the tree satisfying ϕ_T, initial paths which can be of two types:

1. Type 1 paths, representing instantaneous configurations.
2. Type 2 paths, representing transitions between configurations.

In a type 1 path representing an instantaneous configuration (q, w, i) (where w is the contents of the tape and i is the head position), the configuration is coded using atomic propositions from Q and Σ in a straightforward way:

- The first node of the path (i.e. the tree root) bears no symbol – it is used as the "tape left marker".
- If we consider that all initial paths start with index 0 (which is the tree root), then the contents of cell j, say, symbol $w_j = a \in \Sigma$, is an atomic proposition that holds in the jth node of the path.
- Moreover, at each position j along the path only the tape symbol w_j holds. That is, satisfiability of symbols from Σ is mutually exclusive.
- i is the unique position on the path on which the atomic proposition q holds. That is, satisfiability of symbols from Q is also mutually exclusive.
- Whenever a symbol in $Q \cup \Sigma$ holds, the corresponding primed symbol in $Q' \cup \Sigma'$ holds too.
- The whole path contains a position at which $\bot$ holds and from there on it holds forever.
- No symbol from $Q \cup \Sigma$ holds when $\bot$ holds. This codes the finite amount of cell tapes used during simulation.
- At the point where $\bot$ holds, the symbols $\overline{q}$ and $\overline{q}'$ (recall that q is the current state of the Turing Machine). This is needed for coding the connection between type 1 paths and type 2 paths.

In a type 2 path representing a transition between two configurations, say, $(q, w, i) \vdash (r, z, j)$, the unprimed symbols (i.e. symbols from $Q \cup \Sigma$) along the path represent the configuration *before* the transition, while the primed symbols (i.e. symbols from $Q' \cup \Sigma'$) represent the configuration *after* the transition, in a way completely similar to the above description. Hence, we will have some position on the path where symbol q holds, and another position (before or after) where symbol r' holds. Also $\bot$ marks the limit of the available tape space, and $\overline{q}$ and $\overline{r}'$ hold wherever $\bot$ holds.

It then remains to connect type 1 paths (representing instantaneous configurations) with type 2 paths (representing transitions), by means of the observability relations. This connection will be implemented using two agents and their observability relations: one agent being able to see atomic propositions from $Q \cup \overline{Q} \cup \Sigma$, the other seeing $Q' \cup \overline{Q}' \cup \Sigma'$. Formally,

$$AP_1 = Q \cup \overline{Q} \cup \Sigma$$
$$AP_2 = Q' \cup \overline{Q}' \cup \Sigma'$$
$$AP = AP_1 \cup AP_2 \cup \{\bot\}$$

Note that both agents cannot see the value of the symbol $\bot$.

More specifically, we will connect, by means of $\sim_{AP_1}$, each type 1 path representing some configuration (q, w, i) with a type 2 path representing a transition $(q, w, i) \vdash (r, z, j)$. Then, by means of $\sim_{AP_2}$, we code the connection between that type 2 path and another type 1 path, which represents the configuration (r, z, j). The first type of connection is imposed by means of the operator K_1, whereas the second type of connection is ensured by the employment of K_2.

We give in Figure 2 an example of the association between a part of the computation of a Turing machine and a tree. The tree presented in Figure 2 is associated with the following computation of the Turing machine:

$$(q_0, ab, 1) \vdash^{\delta_1} (q_1, cb, 2) \vdash^{\delta_2} (q_2, cBB, 3) \vdash^{\delta_3} (q_3, cBa, 2)$$

where the transitions applied at each step are the following:

$$\delta_1(q_0, a) = (q_1, c, R), \qquad \delta_2(q_1, b) = (q_2, B, R), \qquad \delta_1(q_2, B) = (q_3, a, L)$$

The tree simulates a "guess" that the Turing machine will utilize strictly less than 5 tape cells: on each run there are at most 4 tape cells simulated before the $\perp$ symbol. Here B is the blank tape symbol, whereas R, resp. L denote the commands "move head to the right", resp. "to the left". Note also that the tree node labeled with $\perp, \overline{q}_0, \overline{q}'_0$ is AP_1-similar with the node labeled $\perp, \overline{q}_0, \overline{q}'_1$, which is AP_2-similar with the node labeled $\perp, \overline{q}_1, \overline{q}'_1$. This connection implemented by the composition of $\sim_{AP_1}$ with $\sim_{AP_2}$ encodes the first step in the above computation.

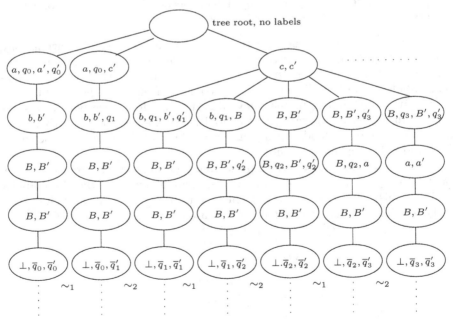

Fig. 2. Simulating a sequence of transitions of a Turing machine within a CTLK tree

Formally, ϕ_T is the conjunction of the following formulas:

1. ϕ_1, specifying that symbols in the same unprimed/primed/overlined set are mutually exclusive:

$$\phi_1 : A\Box \left(\bigwedge_{q,r \in Q, q \neq r} (\neg(q \wedge r) \wedge \neg(q' \wedge r') \wedge \neg(\overline{q} \wedge \overline{r}) \wedge \neg(\overline{q}' \wedge \overline{r}')) \wedge \right.$$
$$\left. \bigwedge_{a,b \in \Sigma, a \neq b} (\neg(a \wedge b) \wedge \neg(a' \wedge b')) \right)$$

2. ϕ_2, specifying that on each inital path there exists a single occurrence of a state symbol, which marks the position of the R/W head:

$$\phi_2 : \left(A\Diamond \bigvee_{q \in Q} q \right) \wedge \left(A\Diamond \bigvee_{q \in Q} q' \right) \wedge \bigwedge_{q \in Q} A\Box \left((q \to A\bigcirc A\Box \neg q) \wedge (q' \to A\bigcirc A\Box \neg q') \right)$$

3. ϕ_3, which, in combination with ϕ_7 and ϕ_9 below, is used to encode the fact that the simulation of T is done on a finite tape, whose end is marked with $\bot$:

$$\phi_3 : A\Diamond \bot \wedge A\Box \left(\bot \to A\Box \left(\bot \wedge \bigwedge_{z \in \Sigma \cup Q} (\neg z \wedge \neg z') \right) \right)$$

4. ϕ_4 which copies the value of the current state of the configuration into the position of the end marker $\bot$ – this is useful for connecting configurations via K_1 and K_2:

$$\phi_4 : A\Box \left(q \to A\Box(\bot \to \overline{q}) \right) \wedge A\Box \left(q' \to A\Box(\bot \to \overline{q}') \right)$$

5. ϕ_5, specifying that on a path of the first type the primed and unprimed symbols are the same:

$$\phi_5 : A\Box \bigwedge_{q \in Q} \left(q \wedge q' \to A\Box \bigwedge_{a \in \Sigma} a \leftrightarrow a' \right) \wedge A\Box \bigwedge_{q \in Q} \left(E\Diamond(q \wedge q') \to \bigwedge_{a \in \Sigma} a \leftrightarrow a' \right)$$

6. ϕ_6, specifying that on a path of the second type, the primed and unprimed symbols are almost everywhere the same, excepting the current position of the R/W head:

$$\phi_6 : A\Box \bigwedge_{(q,a) \in \mathsf{supp}(\delta)} \left(q \wedge a \wedge \neg q' \to A\Box A\bigcirc \bigwedge_{c \in \Sigma} c \leftrightarrow c' \right) \wedge$$
$$\bigwedge_{(q,a) \in \mathsf{supp}(\delta)} \left(E\Diamond E\bigcirc(q \wedge a \wedge \neg q') \to \bigwedge_{c \in \Sigma} c \leftrightarrow c' \right) \wedge$$

7. ϕ_7, specifying that if a path is of type 1 and encodes a configuration in which a certain transition can be applied, then there exists a path of type 2 in which that transition is applied. (The unique transition which will be applied is the subject of ϕ_8.) ϕ_7 also specifies that the "target" type 2 path carries the end marker at the same position as the "source" type 1 path:

$$\phi_7 : A\Box \bigwedge_{(q,a) \in \mathsf{supp}(\delta)} \left(q \wedge a \wedge q' \to A\Box \left(\bot \to K_1 \bot \wedge P_1 \neg \overline{q}' \right) \right)$$

8. ϕ_8, specifying that each transition which can be applied in a certain configuration must be applied in that configuration:

$$\phi_8 : A\square \bigwedge_{q\in Q, a\in\Sigma, \delta(q,a)=(r,b,dir)} (q \wedge a \wedge \neg q' \rightarrow b') \wedge$$

$$A\bigcirc A\square \bigwedge_{q\in Q, a\in\Sigma, \delta(q,a)=(r,b,L)} (E\bigcirc(q \wedge a \wedge \neg q') \rightarrow r') \wedge$$

$$A\square \bigwedge_{q\in Q, a\in\Sigma, \delta(q,a)=(r,b,R)} (q \wedge a \wedge \neg q' \rightarrow A\bigcirc r')$$

Recall that T is deterministic, and hence in each configuration at most one transition can be applied. Note also that the $A\bigcirc$ operator is needed at the beginning of the second line above, in order to code the situations when the head is on the first cell and tries to move left – in such situations the machine halts, no next configuration exists, and therefore our formula ϕ_T needs to be unsatisfiable.

9. ϕ_9, specifying that the outcome of a transition, as coded in a type 2 path, is copied, via $\sim_{AP_2}$, into a type 1 path. ϕ_9 also specifies that the "target" type 1 path carries the end marker at the same position as the "source" type 2 path:

$$\phi_9 : \bigwedge_{q\in Q} A\square \left(q' \wedge \neg q \rightarrow A\square \left(\bot \rightarrow K_2\bot \wedge P_2\overline{q} \right) \right)$$

10. ϕ_{10}, encoding the initial configuration of T:

$$\phi_{10} : \neg\bot \wedge \left(\bigwedge_{z\in Q\cup\Sigma} \neg z \wedge \neg z' \right) \wedge E\bigcirc \left(q_0 \wedge q_0' \wedge B \wedge B' \wedge E\bigcirc \left((B \wedge B') EU \bot \right) \right)$$

Here B represents the blank symbol from Σ. Note that the first position in each path represents the beginning of the tape, hence it does not code any tape cell.

Hence the formula ϕ_T ensures the existence of a path coding the initial configuration of T, together with a guess of the amount of tape space needed for simulating T until it stops or repeats an already visited a configuration. ϕ_T will then ensure, by means of ϕ_7, ϕ_8 and ϕ_9, that once a type 1 path encoding an instantaneous configuration exists in the tree, and from that configuration some transition may be fired, then a type 2 path encoding that transition exists, and the resulting configuration is also encoded in another type 1 path of the tree. It then follows that $(t, \varepsilon) \models \phi_T$ for some tree t if and only if there exists a finite subset Z of inital paths such that Z represents the evolution of the Turing machine T, starting with a blank tape, and either halting in a final state, or re-entering periodically into a configuration – that is, iff the Infinite Visiting Problem has a negative answer for T.   □

4 Model-Checking $CTLK_{prs}$

In this section we present a direct approach to model-checking $CTLK_{prs}$. Our approach is to reutilize the classical state-labeling technique for CTL model-checking from [CES86], by adding a procedure that does state labeling with K_i formulas. This extra procedure requires that the system be "sufficiently expanded" such that each state be labeled with some knowledge formula that holds in the state. This implies the necessity of a subset construction.

An *n-agent system* is a finite representation of an AP-labeled tree. Formally, it is a tuple $\mathcal{A} = (Q, AP_1, \ldots, AP_n, AP_0, \pi, \delta, q_0)$ where Q is a finite set of *states*, $\pi : Q \to 2^{AP}$ is the state labeling (here $AP = \bigcup_{0 \le i \le n} AP_i$), $\delta : Q \to 2^Q$ is the state transition function, and $q_0 \subseteq Q$ is initial state. The n agents are denoted $1, \ldots, n$, and can observe respectively $AP_1, \ldots, AP_n$, whereas AP_0 is not observable by anyone.

We denote $\pi_i : Q \to AP_i$ ($0 \le i \le n$) as the mapping defined by $\pi_i(S) = \pi(S) \cap AP_i$. We also abuse notation and use δ as both a function $\delta : Q \to 2^Q$ and a relation $\delta \subseteq Q \times Q$.

The *tree of behaviors generated by* $\mathcal{A}$ is the tree $t_{\mathcal{A}} : \mathbb{N} \rightarrow Q \times AP$ defined inductively as follows:

1. $\varepsilon \in \mathsf{supp}(t_{\mathcal{A}})$ and $t_{\mathcal{A}}(\varepsilon) = (q_0, \pi(q_0))$.
2. If $x \in \mathsf{supp}(t_{\mathcal{A}})$ with $t_{\mathcal{A}}(x) = (q, P)$ and $card(\delta(q)) = k$, then
   - x has exactly k "sons" in t, i.e. $xi \in \mathsf{supp}(t_{\mathcal{A}})$ iff $1 \le i \le k$
   - there exists a bijection $\sigma : \{1, \ldots, k\} \to \delta(q)$ such that for all $1 \le i \le k$, $t_{\mathcal{A}}(xi) = (\sigma(i), \pi(\sigma(i)))$

Note that $t_{\mathcal{A}}\big|_{AP}$, the projection of the tree generated by $\mathcal{A}$ onto AP, is an AP-labeled tree, and, as such, is a model for $CTLK_{prs}$. We then say that $\mathcal{A}$ is a *model* of a $CTLK_{prs}$ formula ϕ if $(t_{\mathcal{A}}\big|_{AP}, \varepsilon) \models \phi$.

Problem 2 (Model-checking problem for $CTLK_{prs}$). Given an n-agent system $\mathcal{A}$ and a formula ϕ, is $\mathcal{A}$ a model of ϕ?

Theorem 2. *The model-checking problem for $CTLK_{prs}$ is decidable.*

Proof. The technique that we use is to transform $\mathcal{A}$ into another n-agent system $\mathcal{A}'$ generating the same AP-tree (modulo reordering of brother nodes). The new system would have its state space Q' decomposed into states Q'_ψ which satisfy some subformula ψ of ϕ and some states $Q'_{\neg\psi}$ which do not satisfy it. Then we will add a new propositional symbol p_ψ to AP, which will be appended to the labels of Q_ψ (and will not be appended to the labels of $Q_{\neg\psi}$), and iterate the whole procedure by structural induction on ϕ. The new propositional symbol will not be observable by any agent, hence will be appended to AP_0. For the original CTL operators, it will be the case that $\mathcal{A}$ and $\mathcal{A}'$ are the same. Only for the K_i operators we will need to apply a particular subset construction to $\mathcal{A}$ in order to get $\mathcal{A}'$.

So consider first ϕ is in one of the forms $A\bigcirc p$, $p_1 \, A\mathcal{U} \, p_2$, $p_1 \, E\mathcal{U} \, p_2$ or $K_i p$, where p, p_1, p_2 are atomic propositions. The first three constructions presented below are exactly those used for model-checking CTL [CES86], that we re-state here for the sake of self-containment.

For all temporal operators, the main idea is to split the state space into those states that satisfy the formula and those that do not satisfy it. This can be done in linear time, as follows:

For the case of $\phi = A\bigcirc p$, we partition Q in two sets of states:

$$Q_{A\bigcirc p} = \{q \in Q \mid \forall q' \in \delta(q), p \in \pi(q')\}$$
$$Q_{\neg A\bigcirc p} = \{q \in Q \mid \exists q' \in \delta(q), p \notin \pi(q')\}$$

The following lemma shows that our splitting is correct:

Lemma 1. $(t_{\mathcal{A}}\big|_{AP}, x) \models A\bigcirc p$ if and only if for $q = t_{\mathcal{A}}(x)\big|_Q$ we have that $q \in Q_{A\bigcirc p}$.

For the case of $\phi = p_1\,A\mathcal{U}\,p_2$, we partition again Q into two sets of states:

1. $Q_{\neg(p_1\,A\mathcal{U}\,p_2)}$ is the set of states $q \in Q$ for which there exists a subset of states $Q_q \subseteq Q$ such that:
   (a) For all $q' \in Q_q$, $p_1 \in \pi(q)$ and $p_2 \notin \pi(q)$.
   (b) Q_q is strongly connected w.r.t. δ.
   (c) There exists a path $\rho = (q_i)_{1\leq i \leq l}$ connecting q to some state $q' \in Q_q$ such that $p_1 \in \pi(q_i)$ and $p_2 \notin \pi(q_i)$ for all $1 \leq i \leq l$.
2. $Q_{p_1\,A\mathcal{U}\,p_2} = Q \setminus Q_{\neg(p_1\,A\mathcal{U}\,p_2)}$.

Similarly to the first case, we get:

Lemma 2. $(t_{\mathcal{A}}\big|_{AP}, x) \models p_1\,A\mathcal{U}\,p_2$ if and only if for $q = t_{\mathcal{A}}(x)\big|_Q$ we have that $q \in Q_{p_1\,A\mathcal{U}\,p_2}$.

For the case of $\phi = p_1\,E\mathcal{U}\,p_2$, the partition of Q is the following:

1. $Q_{p_1\,E\mathcal{U}\,p_2}$ is the set of states $q \in Q$ for which there exists some state $q' \in Q$ such that
   (a) $p_2 \in \pi(q')$
   (b) There exists a path $\rho = (q_i)_{1\leq i \leq l}$ connecting q to q' such that $p_1 \in \pi(q_i)$ for all $1 \leq i \leq l$.
2. $Q_{\neg(p_1\,E\mathcal{U}\,p_2)} = Q \setminus Q_{p_1\,E\mathcal{U}\,p_2}$.

As above, we also have:

Lemma 3. $(t_{\mathcal{A}}\big|_{AP}, x) \models p_1\,E\mathcal{U}\,p_2$ if and only if for $q = t_{\mathcal{A}}(x)\big|_Q$ we have that $q \in Q_{p_1\,E\mathcal{U}\,p_2}$.

The last construction, for $\phi = K_i p$, no longer labels existing states, but needs to split the states of the given model in order to be able to sufficiently distinguish states that have identical history, as seen by agent i. Therefore, we first build a system which generates the same AP-tree as $\mathcal{A}$, but contains sufficient information about the runs that are identically observable by agent i. Intuitively, the new automaton is an unfolding of the deterministic automaton (without final states) which accepts the same language as $L(\mathcal{A})\big|_{AP_i}$, that is, the projection of the language of $\mathcal{A}$ onto AP_i.

Given a subset $R \subseteq Q$ and a set of atomic propositions $A \subseteq AP$, we denote

$$\delta_A(R) = \{r' \in Q \mid A \subseteq \pi(r') \text{ and } \exists r \in R \text{ with } r' \in \delta(r)\}$$

The new system is $\tilde{\mathcal{A}} = (\tilde{Q}, AP_1, \ldots, AP_n, AP_0, \tilde{\pi}, \tilde{\delta}, \tilde{q}_0)$ where:

$$\tilde{Q} = \{(q, R) \mid R \subseteq Q, \forall r \in R, \pi_i(q) = \pi_i(r)\}$$
$$\tilde{q}_0 = (q_0, \{q_0\})$$

and for all $(q, R) \in \tilde{Q}$,

$$\tilde{\pi}(q, R) = \pi(q)$$
$$\tilde{\delta}(q, R) = \{(q', R') \mid q' \in \delta(q), R' = \delta_{\pi_i(q')}(R)\}$$

The following proposition says that the tree generated by $\tilde{\mathcal{A}}$ and the tree generated by $\mathcal{A}$ are the same, modulo node rearrangement:

Proposition 1. $t_{\tilde{\mathcal{A}}}\big|_{AP} \simeq t_{\mathcal{A}}\big|_{AP}$.

This result is a consequence of the fact that $\tilde{\mathcal{A}}$ is an *in-splitting* of $\mathcal{A}$ [LM95], that is, the mapping $f : \tilde{Q} \to Q$ defined by $f(q, R) = q$ is a surjective mapping for which, for each $(q, R) \in \tilde{Q}$, there exists a bijection between $\delta(f(q, R))$ and $f(\delta(q, R))$, which preserves the AP-labels.

We then partition $\tilde{Q}$ into two sets of states according to whether $K_i p$ holds or not:

$$\tilde{Q}_{K_i p} = \{(q, R) \in \tilde{Q} \mid p \in \pi(q) \text{ and } \forall r \in R, p \in \pi(r)\}$$
$$\tilde{Q}_{\neg K_i p} = \tilde{Q} \setminus Q_{K_i p}$$

Lemma 4. $(t_{\tilde{\mathcal{A}}}\big|_{AP}, x) \models K_i p$ *if and only if for* $(q, R) = t_{\tilde{\mathcal{A}}}(x)\big|_{\tilde{Q}}$ *we have that* $(q, R) \in \tilde{Q}_{K_i p}$.

It then only remains to augment AP in each system such that it includes a new atomic propostion p_ϕ, where ϕ is the formula that was used to partition the state space. Formally, assume that, from the initial system $\mathcal{A}$ and formula ϕ we construct the system $\mathcal{B} = (Q', AP_1, \ldots, AP_n, AP_0, \pi', \delta', q'_0)$ (which in most of the cases above is again $\mathcal{A}$!), and that its state space is partitioned into Q_ϕ and $Q_{\neg\phi}$, with $Q_\phi \cup Q_{\neg\phi} = Q, Q_\phi \cap Q_{\neg\phi} = \emptyset$. We then construct $\mathcal{B}' = (Q', AP_1, \ldots, AP_n, AP'_0, \pi'', \delta', q'_0)$ by augmenting the set of atomic proposition with a new propositional symbol p_ϕ, $AP'_0 = AP_0 \cup \{p_\phi\}$, and relabel accordingly all states: for all $q \in Q'$,

$$\pi''(q') = \pi'(q') \cup \{p_\phi \mid q \in Q_\phi\}$$

The final system $\mathcal{A}_{fin}$ gives the answer to the model-checking problem: $\mathcal{A} \models \phi$ if and only if the initial state in $\mathcal{A}_{fin}$ si labeled with p_ϕ. $\qquad\square$

The complexity of the algorithm is nonelementary, as each knowledge subformula involves a subset construction. It then follows that the complexity of model-checking CTL with knowledge is similar to the complexity of model-checking LTL with knowledge on a synchronous & perfect recall semantics – which, as shown by [vdMS99], is nonelementary. Note that this is in contrast with the classic case, in which model-checking CTL is easier than model-checking LTL [CES86].

5 Conclusions

We have shown that satisfiability is undecidable for CTL extended with knowledge operators, in a synchronous and perfect recall semantics. Our result holds in the absence of common knowledge operators, unlike some other well-known undecidability results on temporal epistemic logics.

We have also given a direct decision procedure for the model-checking problem for $CTLK_{prs}$, which extends the classical algorithm from [CES86]. The algorithm is based on a subset construction which associates with each state q the set of states within the model that give the same observed history for some agent as all runs that reach q.

The undecidability result was obtained while trying to give an automata-based procedure for model-checking CTL. The author tried to build a class of automata for which the "star-free" subclass would be equivalent with formulas in $CTLK_{prs}$. Unfortunately it appeared that no such automata had a decidable emptiness problem. The exact class of automata that would be equivalent with some extension of $CTLK_{prs}$ with "counting" capabilities is still to be identified.

Both our results rely on the synchrony and perfect recall assumptions. It would be interesting to investigate whether any of these can be relaxed, and also whether they can be translated in a non-learning semantics.

The author acknowledges the many discussions with Dimitar Guelev on temporal logics with knowledge, during his visit at the LACL in October-November 2007.

References

[CES86]   Clarke, E.M., Emerson, E.A., Sistla, A.P.: Automatic verification of finite-state concurrent systems using temporal logic specifications. ACM Transactions of Programming Languages and Systems 8(2), 244–263 (1986)

[DW99]   Dickhöfer, M., Wilke, T.: Timed alternating tree automata: The automata-theoretic solution to the TCTL model checking problem. In: Wiedermann, J., Van Emde Boas, P., Nielsen, M. (eds.) ICALP 1999. LNCS, vol. 1644, pp. 281–290. Springer, Heidelberg (1999)

[EKM98]   Elgaard, J., Klarlund, N., Møller, A.: Mona 1.x: New techniques for ws1s and ws2s. In: Y. Vardi, M. (ed.) CAV 1998. LNCS, vol. 1427, pp. 516–520. Springer, Heidelberg (1998)

[ER66]   Elgot, C.C., Rabin, M.O.: Decidability and undecidability of extensions of second (first) order theory of (generalized) successor. Journal of Symbolic Logic 31(2), 169–181 (1966)

[FHV04]   Fagin, R., Halpern, J., Vardi, M.: Reasoning about knowledge. The MIT
          Press, Cambridge (2004)
[HV86]   Halpern, J.Y., Vardi, M.Y.: The complexity of reasoning about knowledge
          and time: Extended abstract. In: Proceedings of STOC 1986, pp. 304–315
          (1986), https://www.cs.rice.edu/~vardi/papers/stoc86r1.pdf.gz
[HV89]   Halpern, J.Y., Vardi, M.Y.: The complexity of reasoning about knowledge
          and time. I. Lower bounds. Journal of Computer System Sciences 38(1),
          195–237 (1989)
[KLN+06] Kacprzak, M., Lomuscio, A., Niewiadomski, A., Penczek, W., Raimondi, F.,
          Szreter, M.: Comparing BDD and SAT based techniques for model check-
          ing Chaum's Dining Cryptographers Protocol. Fundamenta Informaticae
          72(1-3), 215–234 (2006)
[LM95]   Lind, D., Marcus, B.: An Introduction to Symbolic Dynamics and Coding.
          Cambridge University Press, Cambridge (1995)
[LR06]   Lomuscio, A., Raimondi, F.: The complexity of model checking concurrent
          programs against CTLK specifications. In: Baldoni, M., Endriss, U. (eds.)
          DALT 2006. LNCS, vol. 4327, pp. 29–42. Springer, Heidelberg (2006)
[RL05]   Raimondi, F., Lomuscio, A.: Automatic verification of multi-agent systems
          by model checking via ordered binary decision diagrams. Journal of Applied
          Logic 5(2), 235–251 (2005)
[SG02]   Shilov, N.V., Garanina, N.O.: Model checking knowledge and fixpoints.
          In: Proceedings of FICS 2002, Extended version available as Preprint 98,
          pp. 25–39. Ershov Institute of Informatics, Novosibirsk (2002)
[Tho92]   Thomas, W.: Infinite trees and automaton-definable relations over ω-words.
          Theoretical Computer Science 103(1), 143–159 (1992)
[vBP06]   van Benthem, J., Pacuit, E.: The tree of knowledge in action: Towards
          a common perspective. In: Governatori, G., Hodkinson, I.M., Venema, Y.
          (eds.) Proceedings of AiML 2006, pp. 87–106. College Publications (2006)
[vdM98]   van der Meyden, R.: Common knowledge and update in finite environments.
          Information and Computation 140(2), 115–157 (1998)
[vdMS99] van der Meyden, R., Shilov, N.V.: Model checking knowledge and time
          in systems with perfect recall (extended abstract). In: Pandu Rangan,
          C., Raman, V., Ramanujam, R. (eds.) FST TCS 1999. LNCS, vol. 1738,
          pp. 432–445. Springer, Heidelberg (1999)
[vdMS04] van der Meyden, R., Su, K.: Symbolic model checking the knowledge of
          the dining cryptographers. In: Proceedings of the 17th IEEE Computer
          Security Foundations Workshop (CSFW-17 2004), p. 280. IEEE Computer
          Society Press, Los Alamitos (2004)

Contracts Violation: Justification via Argumentation

Stefania Costantini, Arianna Tocchio, and Panagiota Tsintza

Dip. di Informatica, Università di L'Aquila, Coppito 67100, L'Aquila, Italy
`stefcost@di.univaq.it`, `tocchio@di.univaq.it`,
`panagiota.tsintza@di.univaq.it`

Abstract. An important field of application of intelligent logical agents, where rationality plays a main role, is that of automated negotiation. Our work is related to the use of argumentation in the field of negotiation. In particular, we are interested in contract violations, and in the construction of justifications to motivate the violation itself and recover if possible the contract on modified conditions. We propose a temporal modal logic language able to support and depict the arguments/justification used in dialectical disputes and we consider suitable algorithms and mechanisms to introduce and manage justifications.

1 Introduction

Agents living in open societies need to interact in order to decide and coordinate actions that have to be undertaken in order to fulfill a specific and desired goal. In particular, in the case of negotiation (for a survey the reader may refer to [1]), agents have to sign and respect contracts ("deals") that may fail if one of the parties is for some reason unable to fulfil its commitments. In this paper, we intend to present a computational logic framework where an agent can on the one hand try to justify itself in case of a violation by proposing a justification, and on the other hand can evaluate the arguments received by another party that have not respected a deal. The overall objective is that of recovering, if possible, a failed contract on a modified base, even by paying some kind of penalty or performing suitable recovering actions. This objective is of interest not only to single agents, but to the whole "society" where they "live".

In general terms, a successful negotiation process leads to an agreement, called a *contract*, between the involved parties. Typically, a "final" or "signed" contract contains commitments accepted by both parts, that have to be respected in order to conclude the agreement. An omission of a commitment leads to the violation of a signed contract. The omission of a commitment by an agent may have several causes: some of them may depend upon the involved agent, such as lack of resources and/or of capabilities; sometimes however, environmental or social obstructions may occur that the agent is not able to cope with. Therefore, an agent that has violated a contract should be allowed to justify itself by informing the opponent about the particular causes that led to this violation. We propose a formalization for justifications, intended as motivations leading to the omission of a commitment, as a particular kind of argumentations.

Real-world contracts include a number of sanctions and repair actions to be undertaken in case of a contract violation. That is, to each particular case of violation it is in general possible to associate a set of repair rules aimed at trying to recover the contract.

M. Fisher, F. Sadri, and M. Thielscher (Eds.): CLIMA IX, LNAI 5405, pp. 132–153, 2009.

The result of the repair actions can be a new, re-elaborated contract. The violation may however imply an alteration of trust and confidence of each involved agent w.r.t. the opposer. In order to model contract violations and justifications, in this paper we propose a modal/temporal logic (called "CML" for "Contract Major Logic) to represent not only arguments, but also justifications, motivations, lack of resources and lack of capabilities. This new logic, that extends the well-established temporal logic METATEM ([2] [3]), on the one hand is equipped with temporal operators defined on intervals of time, and on the other hand provides specific modal operators for constructing justifications. Intervals are useful because contracts have a date of formulation, a date of expiration, etc. We adapt METATEM rules so as to be able to use them in order to perform run-time controls of contract and violations and, whenever they occur, start an argumentation about justifications, that can lead as a *repair* to a recovered contract.

The reason why we introduce this new logic is that the above-mentioned aspects related to real-world negotiation scenarios cannot easily be expressed in other logics neither directly nor via straightforward definitions. However, this particular application domain is of such practical importance as to motivate the introduction of an "ad hoc" extensions of existing logics. This extension will be in the future also experimented in other related realms involving pacts, violation of pacts, arguments, agreements and disagreements.

We introduce as a background an abstract model for agents and agent societies. The model is abstract in the sense that, in our view, many existing formalisms and architectures for agents might be seen, as they are or with limited enhancements, as instances of this model. A society of agents is considered as a collection of interacting agents that may possibly differ w.r.t. the roles that they play in the society itself. In particular, they will be two classes of negotiating agents, offerers and buyers, and there will also be "institutional" agents controlling deals, and "informer" agents that will distribute relevant information to the other members of the society. Since agents have to interact with the environment, within each agent we assume a multi-layered structure where a special layer (called "IL" for "Interaction Layer") is dedicated to the interaction. This level will be directly linked with a second one (called "GL" for "Ground Layer"), dedicated to the storage and the run-time update of information about other agents of the society and about deals. A third, distinct, layer, the "SL" (for "Sanction Layer") will be dedicated to the storage of sanctions and repair actions.

The main contribution of this paper consists of the definition of the agent model and, most important, of the CML logic and its semantics and use. An operational mechanism is also outlined. In perspective, we intend to propose a comprehensive argumentation-based framework for negotiation centered on CML where violated contracts can be recovered by means of justifications and paying sanctions. Then, in this paper we will discuss at least in general terms all the aspects of this framework, including those that will be treated in detail in subsequent work. This paper is therefore structured as follows. In Section 2, we present a generic layered model to be adopted by the negotiating agents. In Section 3, we define different types of contracts such as generic deals and signed contracts, and we also define contracts violations. In Section 4 we discuss which can be the reasons that can lead agents to omit a commitment. In Section 5, we present the CML language as an extension to the METATEM temporal logic and in Section 6 we discuss in general terms how an agent should detect the occurrence of a violation and how it should react to it. Section 7 treats, with reference to our setting, the problem of safety of interactions, concerning trust and reputation of the involved parties. Section 8 discusses a possible argumentation-based account of the proposed framework. Section 9 is dedicated to both related and future work, and in Section 10 we conclude.

2   Abstract Model for a Society of Negotiating Agents

Negotiating agent are part of a multi-agent system, and "live" in an environment that we call *society*. In order to conclude an agreement, agents have to interact and exchange arguments (or deals). The society is composed of collections of agents, forming different groups according to their assigned role ([4]). A basic group is that of the *Institutions*, including *Institutional agents* offering a service or more generally a negotiation object. It may include banks, travel agencies, e-shops, etc., It will necessarily include some kind of *Governing Structures*, for example the government or a court of law, and *Informer* or *Expert agents*, that offer information such as trust and confidence on others. Finally, the society includes a variety of agents interested in the offered services.

An important role of institutional agents is that of promptly and accurately reporting observations as events occur, thus forming a publicly-available knowledge base that represents the history of the society and the current state-of-affairs. These designated agents posses the "institutionalized power" (a standard feature of any norm-governed system, including agent societies [5,6]) and can be required by the other agents to create specific relations on present or past states of affairs.

Here we propose an abstract layered model to be adopted for modeling agents belonging to a society. First, this model has to contain a component responsible for the interaction between agents that will be called the *IL*, for Interaction-Level, including interaction mechanisms to enforce and regulate the interaction with the society. Then, we introduce another level that we call the Ground-Level (*GL*), containing information concerning other agents. Therefore, the Ground-Level contains information about services and products, about deals and contracts made so far, and about agents of the society -including the agent itself- (for example it may record trust, beliefs and confidence). Part of the information contained in this level has been obtained by means of the *Institutional agents* included in the society. Then, the Ground-Level contains components able to take decisions on the Interaction-Level and coordinate its activities. Besides, the Ground-Level contains a set of arguments and counterarguments that can be used in order to response/react to each interaction. After an interaction has taken place, the Ground-Level decides which modifications have to be undertaken and will update itself to incorporate new information. *IL* and *GL* include a number of components, among which we mention the following.

- *Informational component*, including subcomponents containing *beliefs*, *abilities* and *confidence*. Subcomponent of belief provides modules for reasoning, planning, goal identification, reactivity and proactivity. The component of confidence, instead, is the one responsible of reasoning about the confidence of the agent in its own ability about accomplishing a goal, and can also be affected by the opinion of the informer agents.

- *Constraint component*, that contains all constraints, including temporal ones, useful to validate the dynamic behavior of agents; some constraints will cope with the appropriate performance of actions, and also specify what should happen and what should not happen (for a discussion of this aspect, the reader can see [7]).
- *Motivational component*, that uses a suitable logic for representing agents desires, intention and goals. "Desires" includes all goals adopted while "intentions" includes all goals in execution and under consideration.
- A component including a set of mechanism for managing beliefs, abilities and confidence and their alterations. Here we also include an update mechanism in order to update the informational and motivational component.
- A set of arguments and counterarguments as well as mechanisms for evaluating opponents' arguments and select the most appropriate response to each one.

Contracts result in a violation when at least one of the agents involved in the process either does not fulfill all terms of the contract or does not respects the promises made so far. In real world, each violation of a contract entails sanctions and specific actions that can be seen as "repair actions", in order to "correct" violations. For instance, the delay of a payment can be "repaired" by forcing the agent to pay an extra fee. Of course, not all violations have the same impact: there will be mild breaches or even violations that are not of direct responsibility of the involved agents. Therefore, not all sanctions have to be of the same strictness. Rather, in general a different sanction and/or repair action will be associated to each type of violation (like, e.g., in the HARMON*IA* framework, [8]). This association is stored in a third layer called the Sanction-Level, or *SL*. This level has to be linked with both Ground and Interaction Levels.

3   Contracts: Definitions

In the following, we define in a formal way what a contract is and how it can be violated. Consider two agents involved in a negotiation process via argumentation. We define a general commitment of an agent as follows.

Definition 1 (Deal). *Let Ag be the set of agents and $n \in \mathbb{N}$ be the number of issues used by each agent during the negotiation process. A deal δ_j by an agent j can be defined as the set $\delta_j = \{x_1 : value_1, ..., x_n : value_n\}$, where x_i ($1 \leq i \leq n$) describe all issues that agent j is committed/intending to realize and $value_1, ..., value_n$ their assignment (that can be either numerics or alphabetical).*

A contract violation by an agent occurs in the case where at least one member of the deal set is not fulfilled:

Definition 2 (Disagreement). *Let D be a set of deals and $\delta_j \in D$ a generic deal by agent A then let*

- *$Iss(\delta_j) = \{x_1, ..., x_n\}$ be the mapping from the deal δ_j to the set of issues;*
- *$Value(\delta_j) = \{value_1, ..., value_n\}$ be the mapping from the deal δ_j to the set of assigned values, and*
- *$Obs\_values(\delta_j) = \{value'_1, ..., value'_n\}$ be the set of observed values.*

Then, a disagreement is a tuple (x_i, V_{c_i}, V_{a_i}) *where* $x_i \in Iss(\delta_j)$, $V_{c_i} \in Value(\delta_j)$, $V_{a_i} \in Obs\_values(\delta_j)$ *and occurs when* $V_{c_i} \neq V_{a_i}$.

According to Definition 2, a disagreement occurs when there is some assignment of an issue that does not match with the value observed at the end of the negotiation process.

Definition 3 (Unsigned Contract). *Given two agents, let say A and B, involved in a process of negotiation, then an unsigned contract* δ *is defined as a tuple* $\delta = (\delta_a, \delta_b)$, *where* δ_a *and* δ_b *are deals by agent A and B ,respectively and* $Value(\delta_a) = \{value_{a_1}, ..., value_{a_n}\} \neq Value(\delta_b) = \{value_{b_1}, ..., value_{b_n}\}$

In real-world negotiations, a contract includes a number of sanctions to be undertaken in case one of the parts does not fulfill the commitments. Here, to each contract violation we associate a set R of actions to pursue, called the set of "sanctions" or "repair actions".

Definition 4 (Contract Violation). *Let* D_a *be the set of deals of agent A and* D_b *the set of deals of agent B. Let* $Vial$ *be a set of tuples of the type* $(x_i, value_i, value_i')$ *where* $value_i \neq value_i'$. *Then,* $\ominus : D_a \times D_b \rightarrow Vial$, *called a* difference function, *is the mapping that given two deals returns a vial* ν *which represents the disagreements and that we call a* contract violation.

Consider, for example, two agents negotiating a trip to Rome and let $\delta_a = \{price : 1000, hotel :' 4stars', half-board :' yes'\}$ be the deal of agent A while $\delta_b = \{price : 950, hotel : 3stars, half - board : yes\}$ is the proposal of agent B. Then, the difference function is $\ominus(\{price : 1000, hotel :' 4stars', half - board :' yes'\}, \{price : 950, hotel :' 3stars', half - board :' yes'\}) = \{(price, 1000, 950), (hotel,' 4stars', '3stars')\}$, where $Vial = \{(price, 1000, 950), (hotel,' 4stars',' 3stars')\}$

Definition 5 (Sanction). *Let* D *be a set of deals,* D_a *be the set of deals by agent A and* D_b *the set of deals of agent B and let* $\nu \in Vial$ $(\neq \emptyset)$ *be the set of all violated issues. Then, a* sanction *will be a function of type* $R : Vial \rightarrow D$.

I.e., a sanction recovers a violated contract if it leads to a new deal, presumably including some kind of penalty. For example, assume that agents are negotiating about the price of a trip and both have proposed a deal of 1.000 dollars. At the moment of the payment, agent that buys the trip does not pay the price of the agreement but a lower one, let say 800 dollars. Then, the sanction will be e.g., $R(\{(price, 1000, 800)\}) = (price, 1100)$ where the action of repair is the additional fee of 100 dollars.

Definition 6 (Signed Contract). *Given two agents A and B involved in the process of negotiation, a signed contract* σ *is defined as the couple* (Σ, R) *where* Σ *contains all commitments that agents have to respect for having signed the contract (* $\Sigma = \delta_a \cup \delta_b$ *), while R is the set of sanctions, or repair actions, associated to all issues, that have to be undertaken in case of contract violation.*

4 Contract Violations: Reasons

So far, we have defined types of contracts such as deals, signed or unsigned ones. We intend to formalize contract violations and how they can be repaired by performing repair actions associated to a signed contract. In this section, we discuss the reasons that can lead agents to contract violations. As already mentioned, a contract violation occurs in case at least one of the observed values, associated to a negotiated issue, is different than the value agreed at the time of the contact signature. There are of course different reasons why a contract can be violated. We mention the following.

Lack of resources. In this case, the agent has signed a contract that it cannot respect because it is in lack of the resources needed to fulfill a commitment. For example, an agent can agree to pay a certain price for some kind of service, but at the time of payment it possesses a total amount that is less than the price agreed.

Lack of capabilities. Here an agent, that at time t *believes* to be able to fulfill a plan in order to execute a commitment, realizes, at time $t' > t$ that, due to certain circumstances, it has no more those capabilities. For instance, consider the case in which to fulfill a plan p (let us say a trip to Italy), the agent has first to fulfill a sub-plan p' (drive by car to the airport). Consider then that because of certain changes (e.g., a withdraw of driver's license) the agent realizes to be no more able to execute the sub-plan p', circumstance that forces it to drop also the initial plan p.

Social/Environment factors. These are reasons that lead to a contract violation but do not depend upon agents resources or capabilities but rather upon social factors (someone or something else failed). Let us say that the agent has the resources (money) and the capabilities (drive to the airport) for executing a plan p (e.g., travel to Italy). Consider now that pilots of the air company are on strike just that day. In this case, strike is the social reason that disables the agent to fulfill the contract, and is not of direct responsibility of the involved agent. In this work, we assume that this case can be coped with by resorting to institutional agents: i.e., the external factors that prevented an agent from fulfilling a commitment can be included in a justification if it is possible to certify them by means of an institutional knowledge base.

5 Extending METATEM: Contract Major Logic CML

In this section we introduce the new Contract major Logic (CML) as an extension to the well-known logic METATEM (for a survey on METATEM logic or PML see ([3], [2])). The extension is motivated by the particular kind of application domain we are considering in this paper: in fact, the CML logic is tailored to the realm of contracts and deals. In particular, we add new operators related on the one hand to basic aspects of the agent activity such as beliefs, desires, intentions and knowledge and to the other hand, to "real-life" concepts such as capabilities, (actual) possibility and non-possibility. We introduce these concepts in the logic in order to enable agents to build justifications and try to explain why their obligations could or could not be fulfilled. We also introduce time intervals, starting times and deadlines, as in fact all the above aspects are related to a limited temporal validity, to properties that begin to hold at a certain time but can later expire upon certain conditions, deadlines for fulfilling obligations, etc.

Reasons leading to a contract violation can have their roots both in the dynamic behavior of agents or in the environment (or society) and are related to the events that occur in a dynamic system. Then, we believe it useful to devise for CML an operational mechanisms that allows for run-time monitoring of agents activities. The basic role of dynamic checks is the verification of violations, that have to be coped with by means of some action of *repair*.

5.1 METATEM

In this subsection, we present the basic elements of propositional METATEM logic or PML ([3], [2]). This part can be skipped by a reader who already knows this logic. METATEM is a well-known propositional Linear-time Temporal Logic (LTL) including both future and past operators. LTL logics are called "linear" because, in contrast to Branching Time logics, they that implicitly quantify universally upon all possible paths and evaluate each formula with respect to a vertex-labeled infinite path $p_0p_1...$, where: each vertex p_i in the path corresponds to a point in time (or "time instant" or "state"), p_0 is the present state and each p_i, with $i > 0$, are future states.

First, we present the *syntax* of METATEM. The symbols used by this language are: (i) a set A_C of propositions controlled by the component which is being defined; (ii) a set A_E of propositions controlled by the environment (note that $A_C \cap A_E = \emptyset$); (iii) an alphabet of propositional symbols A_P, obtained as the union of sets A_C and A_E ($A_P = A_C \cup A_E$); (iv) a set of propositional connectives such as **true, false**, $\neg$, $\wedge$, $\vee$, $\Rightarrow$ and $\Leftrightarrow$; (v) a set of temporal connectives; (vi) quantifiers, $\forall$ and $\exists$. The set of temporal connectives is composed of a number of unary and binary connectives referring to future-time and past-time. Given a proposition $p \in A_p$ and the formulae φ and ψ, the syntax of connectives is given below. Note that if φ and ψ are formulae so is their combination.

Unary connectives referring to future and past time:

- $\bigcirc$ that is the "next state" symbol and $\bigcirc\varphi$ stands for: "the formula φ will be true at next state",
- $\square$ that is the "always in future" symbol and $\square\varphi$ means that the formula φ will always be true in the future,
- $\Diamond$ that is the "sometime in future" symbol and $\Diamond\varphi$ stands for: "there is a future state where the formula φ will be true".
- $\bullet$ is the "last state" operator and the formula $\bullet\varphi$ stands for "if there was a last state, then φ was true in that state",
- $\blacklozenge$ is the "some time in past" operator and the formula $\blacklozenge\varphi$ means that formula φ was true in some past state,
- $\blacksquare$ is the "always in the past" and the formula $\blacksquare\varphi$ means that φ was true in all past states,
- $\odot$ is the strong last time operator, where $\odot\varphi \Leftrightarrow \neg\, \bullet\, \neg\varphi$. Note that this state operator can determine the beginning of time by using the formula $\odot$**false**.

Binary connectives referring to future and past time:

- $\mathcal{W}$ is the "unless" (or "weak until") symbol. The formula $\varphi\mathcal{W}\psi$ is true in a state s if the formula ψ is true in a state t, in the future of state s, and φ is true in every state in the time interval $[s,t)$ (t excluded).
- $\mathcal{U}$ is the "strong until" . The formula $\varphi\mathcal{U}\psi$ is true in a state s if the formula ψ is true in a state t which is in the future of state s, and φ is true in every state in the time interval $[s,t]$ (t included). In other worlds, from now on, φ remains true until ψ becomes true.
- $\mathcal{Z}$ is the "zince" (or "weak since") operator. The formula $\varphi\mathcal{Z}\psi$ is true in a state s if the formula ψ is true in a state t (in the past of state s), and φ was true in every state of the time interval $[t,s)$,
- $\mathcal{S}$ that is the "since" operator. The formula $\varphi\mathcal{Z}\psi$ is true in a state s if the formula ψ is true in a state t (in the past of state s), and φ was true in every state of the time interval $[t,s]$. That means that φ was true since ψ was true.

A METATEM program is a set of temporal logic rules in the form:

$$\text{past time antecedent} \rightarrow \text{future time consequent}$$

where the "past time antecedent" is considered as a temporal formula concerning the past while the "future time consequent" is a temporal formula concerning the present and future time. Therefore, a temporal rule is the one determining how the process should progress through stages.

The last part of this section is dedicated to the presentation of METATEM formulae semantics. For doing so, we first define the Model structures used in the interpretation of temporal formulae. In the following, we consider σ to be a *state sequence* $(s_0 s_1 ...)$ and i be the current time instant. We define a *structure* as a pair $(\sigma, i) \in (\mathbb{N} \rightarrow 2^{A_P})$ x $\mathbb{N}$ where A_P is the alphabet of propositional symbols. The relation $\vDash$ is the one giving the interpretation for temporal formulae in the given model structure. In general, a proposition $p \in A_P$ is true in a given model iff it is true in the current moment. As base case, we consider that formula **true** is true in any structure, while **false** is true in no model. Then we have:

Definition 7. Semantics *of temporal connectives is defined as follow:*

- $\sigma, i \vDash$ ***true***
- $\sigma, i \vDash \neg\varphi$ *iff **not** $\sigma, i \vDash \varphi$*
- $\sigma, i \vDash \varphi \wedge \psi$ *iff $\sigma, i \vDash \varphi$ **and** $\sigma, i \vDash \psi$*
- $\sigma, i \vDash \bigcirc \varphi$ *iff $\sigma, i+1 \vDash \varphi$*
- $\sigma, i \vDash \square \varphi$ *iff **forall** $k \in \mathbb{N}$ $\sigma, i+k \vDash \varphi$*
- $\sigma, i \vDash \lozenge \varphi$ *iff **exists some** $k \in \mathbb{N}$ $\sigma, i+k \vDash \varphi$*
- $\sigma, i \vDash \varphi \mathcal{U} \psi$ *iff **exists some** $k \in \mathbb{N}$ such that $\sigma, i+k \vDash \psi$ **and forall** $j \in 0..k-1$, $\sigma, i+j \vDash \varphi$*
- $\sigma, i \vDash \varphi \mathcal{W} \psi$ *iff $\sigma, i \vDash \varphi \mathcal{U} \psi$ or $\sigma, i \vDash \square \varphi$*
- $\sigma, i \vDash \bullet \varphi$ *iff $i > 0$ then $\sigma, i-1 \vDash \varphi$*
- $\sigma, i \vDash \blacksquare \varphi$ *iff **forall** $k \in 1..i$ $\sigma, i-k \vDash \varphi$*
- $\sigma, i \vDash \blacklozenge \varphi$ *iff **exist some** $k \in 1..i$ such that $\sigma, i-k \vDash \varphi$*
- $\sigma, i \vDash \varphi \mathcal{S} \psi$ *iff **exist some** $k \in 1..i$ such that $\sigma, i-k \vDash \psi$ **and forall** $j \in 1..k-1$, $\sigma, i-j \vDash \varphi$*
- $\sigma, i \vDash \varphi \mathcal{Z} \psi$ *iff $\sigma, i \vDash \varphi \mathcal{S} \psi$ or $\sigma, i \vDash \blacksquare \varphi$*

5.2   The Contract Major Logic CML

The basic formulae of the CML language are defined below.

CML propositions are of the type:

$$\varphi ::= \top \mid p_r(\in A_P) \mid \neg\varphi \mid \varphi_1 \wedge \varphi_2 \mid \ldots$$

Actions can be defined in the following way:

$$\alpha ::= atom(\in Action) \mid \alpha; \beta \mid \alpha \vee \beta \mid \alpha^{(n)} \mid \alpha^* \mid !\varphi.$$

Here, $Action$ is the set of atomic actions, $atom$ is an atomic formula representing an action, ; denotes sequencing of actions, $\vee$ the non deterministic choice of actions, $\alpha^{(n)}$ and α^* the bounded and unbounded repetition, respectively, of a generic action α. Finally, $\varphi!$ represents the confirmation of the propositional formula φ.

Given an initial world description, a description of available actions and a goal then a plan can be considered as the sequence of actions that will achieve the goal. Formally, a plan can be defined in the following way:

Definition 8. *Let A_P be the set of observable state variables and O the set of operators. Then a plan p is a tuple $\langle N, b, l \rangle$ where*

- *N is a finite set of nodes*
- *$b \in N$ is the initial node*
- *$l : N \rightarrow (O \times N) \cup 2^{A_P \times N}$ assigns each node*
  - *an operator and a successor node $\langle o, n \rangle \in O \times N$ or*
  - *a set of conditions and successor nodes $\langle \varphi, n \rangle$, where $n \in N$ and φ is a formula over A_P*

Sometimes, by abuse of notation we will consider a plan simply as a set of actions.

After giving the basic definition of propositions and actions in our model, we propose an extension to the METATEM logic, and therefore we add new logical operators. In the setting of negotiation, contracts and deals one often needs to express the interval of time where properties are supposed to hold, and to easily consider future states which are beyond the next state. Then, our first step is to extend METATEM operators to include this feature.

- τ, where the proposition $\tau(s_i)$ is true if s_i is the current state, i.e., we introduce the possibility of accessing the current state;
- $\bigcirc_m$ stands for true in a certain future state, i.e., $\bigcirc_m \varphi$ means that φ should be true at state s_m;
- $\Diamond_m$ stands for bounded eventually, i.e., $\Diamond_m \varphi$ means that φ eventually has to hold somewhere along the path from the current state to s_m;
- $\square_{m,n}$ stands for always in a given interval, i.e., $\square_{m,n}\varphi$ means that φ should become true at most at state s_m and then hold at least until state s_n;
- $\square_{\langle m,n \rangle}$ means that φ should become true just in s_m and then hold until state s_n, and not in s_{n+1}, where nothing is said for the remaining states;
- N stands for "never", i.e., $N\varphi$ means that φ should not become true in any future state;

- $N_{m,n}$ stands for "bounded never", i.e. $N_{m,n}\varphi$ means that φ should not be true in any state between s_m and s_n, included

Then, we introduce, again with reference to time intervals, operators expressing the basic mental attitudes of an agent.

- $K_{m,n}$: K is used to represent agents' knowledge, and $K_{m,n}\varphi$ stands for "agent knows that φ in any state between s_m and s_n";
- K^n is used to represent agents' knowledge and $K^n\varphi$ stands for: "agent knows that φ in state n"
- $D_{m,n}\varphi$ stands for agent's desire that φ be true during states s_m and s_n;
- $B_{m,n}\varphi$ represents the agent's belief of φ being true in the interval s_m,s_n.

Finally, we introduce (again over intervals) operators which express more elaborated (we will say for short "advanced") mental attitudes related to what an agent assumes to be able and not able to do. Thus, advanced attitudes are related to the agent willingness to accept a deal and to its ability to construct a justification in case it cannot fulfil the obligations.

- $Ab_{m,n}(a)$ means that agent has the ability to perform action a in any state of the interval s_m-s_n; it can be a basic agent belief, or the conclusion of having this ability must be somehow derived, in which case it may in turn depend upon the ability of performing other actions (e.g., I am able to take the bus if I have a ticket, I have a ticket if I buy a ticket, and I have the ability of buying a ticket if I have the money, etc).
- $FPoss_{m,n}(a)$ means that agent has the actual possibility of performing action a within states s_m and s_n. This operator can be formalized as $FPoss_{m,n}(a) = \Box_{m,n}(Ab_{m,n}(a))$. In a more general case, where one needs to consider a plan and not a single action, this operator has the form
  $FPoss_{m,n}(a_1;\ldots;a_n) = FPoss_{m,n}(a_1) \wedge \ldots \wedge FPoss_{m,n}(a_n)$
  where the sequence $a_1;\ldots;a_n$ represents a plan p.
- $FCan_{m,n}(a) = B_{m,n}(FPoss_{m,n}(a))$: the agent believes that the execution of action a is possible in the interval and that it is able to execute it; it generalizes to a plan p as before;
- $FCannot_{m,n}(a) = K_{m,n}(\neg FPoss_{m,n}(a))$: the agent knows that the execution of action a is not possible and therefore it is not able to execute it; it generalizes to a plan p where $FCannot_{m,n}(p) = \exists a \in p$ such that $FCannot_{m,n}(a)$.
- $Goal_{m,n}(p) = \neg p \wedge D_{m,n}(p)$: a goal (taking the form of a plan p) not yet fulfilled is still desired.
- $Intend_{m,n}(p) = FCan_{m,n}(p) \wedge Goal_{m,n}(p)$: agent intends to accomplish a goal (taking the form of a plan p) that is still desired can be pursued.
- $Drop_{m,n}(p) = Intend_{m',n'}(p) \wedge FCannot_{m,n}(p)$, where $m' < m$. I.e., an agent drops a plan p if at time m' it had the intention to fulfill the plan but at time $m > m'$ it realizes that it cannot do so anymore because some action of the plan cannot be performed.

The past operators are, symmetrically, the following:

- $\bullet_m$, i.e., given the current state s_i then φ was true at state s_m, with $m < i$;
- $\blacksquare_{m,n}$ is "always in the past" operator stating that, if the current state is s_i and $m \leq n \leq i$ then φ was true in the entire time interval m, n, i.e., the formula $\blacksquare_{m,n}\varphi$ means that φ was true at state s_m and then has held at least until state s_n;
- $\blacksquare_{\langle m,n \rangle}$ is the strict version of $\blacksquare_{m,n}$ where φ was true only in the time interval m, n, i.e., the formula $\blacksquare_{\langle m,n \rangle}$ means that φ became true just in state s_m and then has held just until state s_n.

As before, we can define past operators for knowledge, desires, intentions, abilities, i.e., $PPoss$, $PCan$, $PCannot$ by considering the current state and suitable state intervals.

The semantics of basic CML operators is reported below, as a straightforward extension of the METATEM one.

Definition 9. (Semantics of CML formulae) *Let σ be a state sequence $s_0 s_1 ...$, i the current moment in time, φ, ψ CML-formulae, i the current time instant and m, n generic instances of time with $m \leq i \leq n$. The semantics of CML is defined as:*

- *all basic METATEM operators are defined as in Definition 7;*
- $\sigma, i \vDash \tau(s_0)$, *where* $s_0 \equiv \bullet false$;
- $\sigma, i \vDash \bigcirc_m\varphi$ *iff* $\sigma, m \vDash \varphi$;
- $\sigma, i \vDash \Diamond_m\varphi$ *iff* ***exist some*** $j, j \leq m$: $\sigma, j \vDash \varphi$;
- $\sigma, i \vDash \Box_{m,n}\varphi$ *iff* ***forall*** $m \leq j \leq n$: $\sigma, j \vDash \varphi$;
- $\sigma, i \vDash \Box_{\langle m,n \rangle}\varphi$ *iff* ***forall*** $j, m \leq j \leq n$: $\sigma, j \vDash \varphi$ *and* ***forall*** r: $r < m$: $\sigma, r \vDash \neg\varphi$ *and* $\sigma, n+1 \vDash \neg\varphi$;
- $\sigma, i \vDash N\varphi$ *iff* ***forall*** $j, j \geq 0$: $\sigma, j \vDash \neg\Diamond\varphi$;
- $\sigma, i \vDash K_{m,n}\varphi$, $\sigma \vDash B_{m,n}\varphi$, $\sigma \vDash D_{m,n}\varphi$ *are defined as in modal logic considering that proposition φ has to be true in time interval between states s_m and s_n;*
- $\sigma, i \vDash K^n\varphi$ *iff* $\sigma, n \vDash\varphi$;
- $\sigma, i \vDash K_{m,n}\varphi$;
- $\sigma, i \vDash \bullet_m\varphi$ *iff for* $m < i$: $\sigma, m \vDash\varphi$;
- $\sigma, i \vDash \blacksquare_{m,n}\varphi$ *iff* ***forall*** $j, m \leq j \leq n \leq i$: $\sigma, j \vDash\varphi$;
- $\sigma, i \vDash \blacksquare_{\langle m,n \rangle}\varphi$ *iff* ***forall*** $m \leq j \leq n \leq i$: *then* $\sigma, j \vDash\varphi$ *and* ***forall*** r: $r < m$ *then* $\sigma, r \vDash \neg\varphi$ *and* $\sigma, n+1 \vDash \neg\varphi$;

The semantics of basic advanced mental attitudes can be then derived. It is however important to notice that the "intended" semantics is to some extent *subjective* and *context-dependent*. Consider for instance the operator $FPoss_{m,n}(a)$ by which an agent says to be able to execute action a in a given time interval m, n. Something can come up later that invalidates this assumption. Then, the agent can propose as a justification in case the unfeasibility of a leads to violating some kind of obligation: the opponent agent in the course of an argumentation process can however either accept this justification or not.

We need a further extension for defining state subsequences. In fact, verification of properties will not realistically occur at every state but, rather, at a frequency associated to each property. In such a way, a crucial property for agents' evolution can, e.g., be tested more often than a less relevant one.

Definition 10. *Let σ be an infinite sequence of states $s_0, s_1, \ldots$ of a system. Then, σ^k is the subsequence $s_0, s_{k_1}, s_{k_2}, \ldots$ where for each k_r ($r \geq 1$), $k_r \bmod k = 0$, i.e., $k_r = g \times k$ for some g.*

From previous definition it follows that $\sigma^1 = \sigma, \sigma^2 = s_0, s_2, s_4, \ldots$ and so on: therefore, all the operators introduced above can be redefined for subsequences.

Definition 11. *Let O_p be any of the operators introduced in CML and $k \in \mathbb{N}$ with $k > 1$. Then, O_{p^k} is an operator whose semantics is a variation of the semantics of O_p where each sequence σ_s is replaced by the subsequence $\sigma_s{}^k$.*

5.3 Justification and Motivations

We define here a generic operator called *Motivation* in order to represent the reasons that an agent believes to be at the basis of a contract violation and that the agent can use for proposing a justification to the opponent. Let σ be a signed contract and ν a violation.

Definition 12. *We define Motivation(δ, ν) as a conjunction of CML formulas.*

In this way, an agent can justify itself by informing the opposer that it is, e.g., in lack of resources, knowledge, intentions, capabilities etc. and therefore it is not capable anymore to fulfill a plan or action. A *Motivation* can be an expression such as for instance the following:

$$D_{m,n}(a) \wedge Ab_{m,n}(a) \wedge \tau(s_{n+i}) \wedge \neg Ab_{m',m'+i}(a).$$

In this case, the motivation of the agent is that even if in the past it had the ability and the desire to perform an action a during the interval m–n, after the contract has been signed it realized at state $m' > m$ that it has no more the ability to do so.

Justifications can be formulated as follows:

Definition 13. *Let Motivation be defined as above, $i, j > 0$ and s_m–s_n the interval in which the process of negotiation has taken place. Let φ and φ_C be any formulas. Then, the operator of justification can be formulated as follows:*

$$Justification(\delta, \nu) = \tau(s_{n+i}) \wedge B_{m,n}\varphi \wedge Motivation(\delta, \nu) \wedge K^j\varphi_V$$

where φ represents the premises on the basis of which the agent proposes the justification, and φ_V represents the obligations that the agent understands to have violated, i.e., a transposition of (some elements of) ν into a logical formula.

According to Definition 13, an agent justifies itself in the current state (s_{n+i}), if it has signed a contract in the past on the basis of a belief φ of its ability to fulfil the obligations and has some motivation justifying which action, plan or sub-plans it was not able to complete.

An example of *Justification* is, e.g.,

$$\tau(s_6) \wedge B_{1,3}(D_{1,3}(a) \wedge FPoss_{1,3}(a)) \wedge FCannot_{4,6}(a) \wedge K^6\neg payed(f)).$$

In such a way, the agent affirms that: during stages of negotiation s_1–s_3, it desired and believed it had the possibility to execute action a (e.g., a bank transfer). Later however, it realized at state s_6 that it was no more possible to perform the action, and therefore has reached the certainty not to have payed the required fee f.

5.4   CML Rules

In this section, we propose an operational interpretation of CML rules that can be in principle adopted in any computational logic language based on logic programming. Then, we assume that formulae of the logic language underlying CML are expressed in logic programming. We build upon our related work presented in [9] and [7]. The reasons why we do not adopt for CML the analogous of METATEM temporal rules are at least the following reasons.

- We choose not to exploit the full potential of modal logic in order not to cope with its computational complexity: we aim at keeping its conceptual advantages while staying within a reasonable efficiency.
- We have designed CML so that it can be "incorporated" into any computational logic agent-oriented language/formalism. Then, we propose an operational mechanism which should be compatible with most existing frameworks.
- We wish to leave it open how and when respect/violation of a contract should be verified.

In our operational mechanism, CML rules are attempted periodically, at a certain frequency that can be associated (as a form of control information) to the rule itself: presumably, rules corresponding to more crucial properties will be attempted more often than less relevant ones. For modeling the frequency, we consider *state sequence* and *subsequence* as in Definition 10. Then a CML *rule* is defined as follows.

Definition 14. *A CML rule ρ is a writing of the form $\alpha \;:\; \beta$ or simply β where β is a conjunction including either logic programming literals or CML operators (possibly negated) and α is an atom of the form $p(t_1, \ldots, t_n)$, called the rule representative.*

We now define a "concrete" syntax for CML operators, that consists by considering the operator as a new predicate, and the time interval/threshold and the frequency as arguments of this predicate. Then, a generic operation is defined as follows.

Definition 15. *Let φ be a sentence and OP any of the operators introduced in CML. Then a generic operation will be in the form $OP(m, n; k)\varphi$, where m, n is the time interval and $k \geq 0$ is used to indicate at what frequency a certain sentence has to be verified.*

Note that we can omit frequency if not relevant, resorting to a "default" one. Consider for instance the operation $FPoss(m, n; k)$: this is the operation of "future possibility" and is associated to the CML operator $OP^k = FPoss^k_{m,n}$. In a similar way, this can be done for all CML operators. We want to allow for CML rules where the arguments are variables, whose instantiation will depend on each particular context in which the sentence will be evaluated. Therefore, we define a contextual rule as a rule depending and grounded within an evaluating context χ.

Definition 16. *Let $OP(m, n; k)\varphi$ be a CML rule. The corresponding* contextual CML rule *is a rule of the form $OP(M, N; K)\varphi :: \chi$ where*

- *χ is called the* evaluation context *of the rule, and consists of a conjunction of logic programming literals;*

- M, N, K, *that can be either variables (suitable instantiated by the context χ) or ground terms;*
- *in case that M, N, K are variables, they have to occur in an atom (non-negated literal) of the context χ.*

In order to define the analogous of a METATEM rule, we define CML contextual rules with a consequent, that will be executed whenever the antecedent is verified. The consequent however in this setting is not a generic temporal expression. Rather, it is intended as expressing a *repair* w.r.t. the violation of a contract by the agent itself or by another agent of the society. In fact, a crucial point of our approach consists in specifying the repair actions that an agent has to perform in order to "pay" a sanction for the violation of a signed contract. In the following, a general *rule with repair* is defined. Via these rules, particular violations are linked to the related repair actions.

Definition 17. *Considering the generic formulation of a METATEM rule then a CML rule with a repair is a rule of the form: $OP(M, N; K)\varphi :: \chi \Rightarrow \psi$, where:*

- $OP(M, N; K)\varphi :: \chi$ *is a contextual CML rule;*
- ψ *is called the* repair *action of the rule, and is an expression including either a justification or a motivation (as defined above).*

Operationally, if the left-hand-side is verified, then the right-hand-side is executed. This means, the repair is performed and this should restore the agent to a safe state. The repair is defined as an arbitrary expression encompassing a justification/motivation. However, it should imply at least to send a justification/motivation to the opponent, and possibly also to an institutional agent which is supposed to be in charge to act as a referee. The repair may also possibly imply the payment of a sanction, if associated in advance to the contract. Finally, a repair may start a "dispute" if the opponent agent does not accept the justification/motivation and chooses to react, or "counterattack".

6   Verification of Violations

In our view, when a contract violation occurs (i.e., at least one of the issues included in a signed contract is not respected) deals and motivations/justifications should be broadcasted to all the agents in the society, i.e., should be made publicly available. In fact, the fulfillment of a commitment, or its violation, will determine measurable changes in the environment of some of the agents involved in the negotiation. Also, the validity of a justification can often be verified either by the opponent agents or by third parties. This verification should come prior than the decision to accept/reject this justification. In general, every part involved in a contract is potentially able to violate its contract and exhibit a justification.

Violations should be somehow "rated", as there are violations that are of greater importance than others. Violations of a certain importance should be signalled to the governing structure, that should take the role of third party. The reader may notice a similarity with the human society organization: in fact, establishing such a similarity not only makes it easier to cope with practical cases by encoding as background knowledge the "corpus" of past (human) experience about negotiation, but should also make the system more acceptable by human users. Expert agents have to update their information on the agent that has violated a contract and then be able to inform other agent on future negotiating processes. We are considering the possibility that this type of information can be a particular type of a trust data, as better discussed below. Also, agents are able to check not only if a violation occurred by the other parts participating in the process, but also can control their own behavior and decide how to react when they realize to be no longer able to fulfil an obligation, e.g., in the following different ways:

– by sending a message informing all opposers, that includes a motivation;
– by internally repairing the violation;
– by deciding not to react but rather wait for the opposers to verify the violation and undertake some action.

In case the opponent agent is informed about a violation and receives a justification, it can decide either to accept it or to reject it. This should require a *reason*, and can lead to an argument where each agent counterattacks the other one's positions. The dispute can end either with a new contract or with a broken contact, i.e., with a failed negotiation.

7   Trust and Reputation in Negotiation

In the kind of setting we are considering, trust among the parties and reputation of the involved parties assumes an important role. Namely, "reputation" can be seen the opinion (or, better, a social evaluation) of a group about something, while trust can be seen as a "subjective" expectation of an entity that another entity will perform certain actions. The reader may refer, e.g., to [10,11] and to [12] for a discussion about recent computational models of trust and reputation. Trust is a crucial issue in multi-agents systems and societies and consequently in negotiation and in particular in e-commerce, where trust has been widely studied, e.g., in [13], [14], [15], [16] and many others. The point about an agent having to make a trust decision is that it represents a willingness to expose itself to the risk that something will go wrong. Successful conclusion of an interaction can lead to an increase of the level of trust to the opponent agent. An occurred violation, instead, and a consequent decision not to accept the related justification (if any) can result in a decrease of the level of trust. Then, trust and reputation are especially critical in the kind of negotiation that we are considering, which implies proposing/accepting/rejecting justifications. Our approach, which tries to recover a deal by means of justifications, can be seen as related to the approach of [17], where in fact trust is explicitly considered. In [17], agents try to "persuade" the others by means os suggestive, or "rhetorical" arguments whose role is to lead the negotiation opponent to accept proposals more readily, thus improving the efficiency of the process. In our setting, an agent will be more open to accept a justification if coming by a trusted party with a good reputation, which has behaved well in past interactions. Also, among the factors that may determine what justifications to send and what penalties to accept are the desirability of the deal and the degree of trust on the other party. Therefore, a future direction of our work involves formally introducing in CML an explicit aspect of trust and reputation evaluation where, up to now, we have only made practical experiments with a simple methodology whose motivation and essential functioning are shortly summarized below.

As a matter of fact, we may notice that many approaches, especially in e-commerce, resort to a mediator agent to identify trustworthy entities, where the mediator agent is understood as an institutional agent (this approach is referred to in [18] as "institutionalized trust"). We may however also notice that this is not sufficient, because trusting decisions can be only partly based on the recommendation of a third party, since personal experience and observations of previous interactions necessarily influence an agent's trust decisions. As trust is related to the subjective perception of risk, even if two entities get the same information they might not draw the same conclusions. The same agent can possibly associate different degrees of risk (based on its own observations), and then require/associate different levels of trustworthiness, to different topics. A mediator/directory agent should, however, be notified of the updates of the level of trust performed by agents. The directory agent will then employ suitable algorithms to assess the past behavior of agents, so as to allow avoidance of untrustworthy agents in future.

How the mediator should compute the reputation of agents is a critical issue: in fact, it is important to avoid that the reputation of agents with a basically reliable "history" be spoiled by reports produced by accidentally wrong interactions. Also, the trust level should be associated to context information such as the relative importance of various kinds of violations and the frequency at which they have occurred. As a simple example: an occasional delay cannot spoil a reputation, while repeated delays may lead to consider an agent as unreliable as far as punctuality is concerned. The trust level should moreover be updated after a sufficient number of reports by trustworthy agents.

An important issue in our framework, where we try to recover contract violations by means of justifications is how a loss of trust can be "repaired" (see e.g., [19]). In fact, an offence, namely a contract violation, may be unintentional or exceptional, performed by a benevolent member of the community. Therefore, if a sufficient penalty has been paid, the original level of trust and reputation of the "offender" agent should be later restored. We have defined and implemented [20] a simple form of "forgiveness" supported by three different evaluating systems: a reputation system, a reputation system with a built-in apology forum that may display the offender's apology to the "victim" and a reputation system with a built-in apology forum that also includes a "forgiveness" component. The implementation has been developed using the agent-oriented logic programming language DALI [21,22].

8   Counter Attacks

Argumentation has since long (cf. [23,24]) been advocated as a suitable procedural mechanism for non-trivial forms of negotiation, and has been applied in this context in many ways (for a survey, cf. e.g., [1]). In fact, (cf. [24]) "While proposals, counterproposals and the ability to accept and reject proposals are minimal requirements of negotiation, they are inadequate if agents seek to justify their stance or to persuade others to change their negotiation stance. If there are compelling reasons for adopting a particular position, agents may use arguments to try to change their opponents view, and to make their own proposal more attractive by providing additional meta-level information in its support. The nature and types of arguments are varied, but include threats, rewards, appeals, and so on. Thus, argumentation is a type of negotiation that includes meta-level exchanges to support a negotiators position." The point of view that we assume in this paper, i.e., trying to recover a contract from violations by resorting to some kind of justification, naturally fits into this view and in fact we are working to the definition of a suitable argumentation system for our logic that we try to outline here even though it is still work-in-progress.

Whenever a violation of a contract and a justification occurs, an agent may decide to counterattack all or part of proposed motivations. The decision of counterattacking a motivation can be taken, e.g., by considering a dispute as a derivation as described in [25]. In particular, interactions can be defined as dialogues between agents, that in our case are the agent that violated a contract (possibly, it can be an institutional agent which enters a dispute on behalf of one of its clients) and the opponent one, trying to establish the acceptability of respective beliefs. Possible formulations of these dialogues differ in the level of skepticism of the proponent agent. In GB (Ground-Based) disputes, the agent is completely skeptical in the presence of alternatives; in AB (Admissible-Based) disputes, the main goal of an agent is to find and use any alternative in order to counterattack arguments without attacking itself (and this is our case); in IB (Ideal-Based) disputes, the agent uses only the common parts of alternatives.

In [25], a framework is defined as a tuple $\langle \mathcal{L}_c, \mathcal{R}, \mathcal{A}, \mathcal{C}on \rangle$ where $\mathcal{L}_c$ is the language, $\mathcal{R}$ is the set of the inference rules, $\mathcal{A} \subseteq \mathcal{L}_c$ is an non empty set whose elements are referred to as assumptions and $\mathcal{C}on$ is the total mapping from assumptions in $\mathcal{A}$ into a set of sentences in $\mathcal{L}_c$. Then, the couple $(\mathcal{L}_c, \mathcal{R})$ is the deductive system while the notion of attack between sets of assumptions is defined as follows:

Definition 18. *A set of assumptions X attacks a set of assumptions Y iff there is an argument in favor of some x supported by a subset of X, where $x \in \mathcal{C}on(y)$ and y is in Y.*

A generalized AB-dispute derivation of a defense set A for a sentence φ is a finite sequence of quadruples:

$$\langle \mathcal{P}_0, \mathcal{O}_0, A_0, \mathcal{C}_0 \rangle, \ldots, \langle \mathcal{P}_i, \mathcal{O}_i, A_i, \mathcal{C}_i \rangle, \ldots, \langle \mathcal{P}_n, \mathcal{O}_n, A_n, \mathcal{C}_n \rangle$$

where: $\mathcal{P}_j, \mathcal{O}_j$ ($0 \leq j \leq n$) represents the set of sentences held respectively by the proponent and opponent agents, A_j is the set of assumptions generated by the institutional agent and $\mathcal{C}_n$ is the set of assumptions generated by the agent that the opponent has chosen to counterattack. The derivation algorithm is presented in [25].

Then, the nature of counterattacks and its choice will depend upon the particular operator used to formalize a motivation. For instance, if the received justification is formed in terms of operator K, then the receiving agent can attack arguments of the motivation based on the basic common knowledge ("I know that you know..."). In case, e.g., of motivations formed in terms of $\mathcal{C}annot$, the receiving agent may attack the arguments based, for example, on references received by the society (the *informing agents*) and so on.

9   Related and Future Work

Temporal, modal and deontic logic have been widely used in knowledge representation and automated reasoning and in the formulation of languages, protocols and exchange messages for argumentation-based negotiation.

Modal logic has been used in [26] to build the framework "Commitment and Argument Network" (CAN) in order to represent social commitments, actions that agents associate to these social commitments and arguments that agents use to support their actions. In [27], instead, temporal logic is used to represent contracts and violations, where however the control on the validity of sentences is performed in advance (similarly to model checking [28]) and not dynamically like in the proposed where the use of interval temporal logic allows for run-time checks.

As concerns verification of contracts, [29] introduces a deontic (logic based) policy language, referred to as DPL, to describe enterprise policies. In their approach, an interpreter is capable of using DPL statements to monitor contract execution by detecting contract violations at run-time. Run time verification of properties expressed in linear time temporal logic (LTL) or in timed linear time temporal logic (TLTL) is considered in different works in literature. Here we mention [30], where a three-valued semantics as well as a simple monitor generation procedure is introduced. In [31], the BCL language is presented as an extension of deontic logic to represent contracts, violations and repair actions. Some authors consider violations as special kinds of exceptions where they use these exception to raise conditions in order to repair the violation in the context of contract monitoring (cf., e.g., [32] and [33]). In [34], agents follow the BDI model and are provided with goals and expectations that have to be fulfilled by composing different contracts. Repairs in case of violation are however not provided. A general framework for verifiable agent dialogues is introduced in [35] where the "Multi-Agent Protocol" (MAP) language expresses dialogues in Multi-Agent Systems by defining the patterns of message exchanges that occur between the agents, so as to be independent of the actual rational processes and message-content.

As concerns argumentation-based negotiation, [36] reports various logic languages introduced to this aim and [37] proposes a framework which provides a formal model of argumentation-based reasoning and negotiation, which ensures a clear link between the model and its practical instantiation to belief-desire-intention agents. Among the most recent approaches we have already mentioned [17]. It is also worth mentioning interesting recent work presented at CLIMA-IX: it proposes [38] an argumentation -based framework for modeling contract dispute resolution as a two-level reasoning process, where at the "object level" the acceptability of beliefs and facts is assessed, while at the meta-level the legal doctrines will determine the actual agent's behavior. Agents are equipped with beliefs, goals and argumentation-based decision making mechanisms taking in consideration uncertainties. In [39], the approach is applied to a real-world case-study, so as to demonstrate its effectiveness. This is closely related work, and one future aim of the present work is that of seeking an integration.

A formal abstract model of disputes based on argumentation is provided in [40], which captures both the logical and the procedural aspects of argumentation processes. Here, argumentation protocols, also called rules of order, describe which particular speech-act is acceptable, or legal, in a particular case of dispute. In this framework violations are tolerated but the model allows participating agents to object to illegal actions. The interest in this work is that the general formal definition of argument systems does not actually depend upon any particular logic of disputation. Such a logic (be it non-monotonic or not, defined in terms of arguments or not) is one of the parameters that need to be instantiated when specific argument systems are built. Therefore, as future work we intend to explore the possibility of embedding the model/temporal logic we have proposed here into this framework.

An alternative though intriguing approach is that of [41,42] which consists in a theoretical and computational framework for the executable specification of open agent societies, that are considered as instances of normative systems. The (executable) specification is formalized in two action languages, the C+ language and the Event Calculus, for which inference engines are actually available.

The EU Project CONTRACT adopts in some sense a similar view. In fact, within this project a logical representation of contracts has been proposed different from the one commonly found in the negotiation literature and also adopted in this paper (cf. the references in the project web site [43], e.g., [44]). They assume that a contract is made up of various descriptive elements, for example, stating which ontologies may be used to explain the terms found within it. Most importantly, a contract includes a set of clauses, each of which represents a norm, where can be interpreted as socially derived prescriptions specifying that some set of agents (the norms targets) may, or must, perform some action, or see that some state of affairs occurs. Norms can be understood as regulating the behavior of agents. We believe that this approach will have to be carefully considered in future work.

10   Conclusions

The contribution of this work is twofold: on the one hand, we have formalized justifications and motivations in terms of a new modal/temporal interval logic; on the other hand, we have proposed forms of run-time control, by means of axioms to be dynamically verified where the verification may imply the execution of repair actions. The proposed approach has to be completed by formally defining what is a repair and how it should be managed, including starting and carrying on arguments.

We also mean to explore the choice of counterattacks in our setting, e.g., by representing arguments as trees. Nodes of such trees may contain argumentations, and therefore also justifications, while the ramification represents the different motivations. In such a way, a counterattack to a justification can be reduced to the problem of responding with a counterargument. An envisaged extension to the proposed approach is the definition of algorithms and mechanisms aimed at responding in the most appropriate way to a motivation.In addition, future work includes the implementation of meaningful, and real word, case-studies in order to discover how our approach is behaves when put at work.

References

1. Rahwan, I., Ramchurn, S.D., Jennings, N.R., Peter McBurney, S.P., Sonenberg, L.: Argumentation-based negotiation. Knowledge Engineering Review (2004)
2. Barringer, H., Fisher, M., Gabbay, D.M., Gough, G., Owens, R.: METATEM: A Framework for Programming in Temporal Logic. In: Proceedings of Stepwise Refinement of Distributed Systems, Models, Formalisms, Correctness, REX Workshop, London, UK, pp. 94–129. Springer, Heidelberg (1990)
3. Fisher, M.: Metatem: The story so far. In: Bordini, R.H., Dastani, M., Dix, J., El Fallah Seghrouchni, A. (eds.) PROMAS 2005. LNCS, vol. 3862, pp. 3–22. Springer, Heidelberg (2006)
4. Rodríguez-Aguilar, J.: On the Design and Construction of Agent mediated Electronic Institutions. PhD thesis, Institut d'Investigaciò en Intel.ligència Artificial (IIIA) (2001)
5. Makinson, D.: On the formal representation of rights relations. Journal of Philosophical Logic 15, 403–425 (1986)
6. Jones, A., Sergot, M.: A formal characterisation of institutionalised power. Logic Journal of IGPL 4(3), 427–443 (1996)
7. Costantini, S., Acqua, P.D., Pereira, L.M.: Specification and Dynamic Verification of Agent Properties. In: Proceedings of CLIMA-IX, Int. Worksh. on Computational Logic in Multi-Agent Systems
8. Vázquez-Salceda, J.: The role of Norms and Electronic Institutions in MultiAgent Systems applied to complex domains. The HARMONIA framework. PhD thesis, Artificial Intelligence PhD. Program, Universitat Politècnica de Catalunya (April 2003)
9. Costantini, S., Acqua, P.D., Pereira, L.M.: A Multi-layer Framework for Evolving and Learning Agents. In: Cox, M.T., Raja, A. (eds.) Proceedings of Metareasoning, Thinking about thinking workshop at AAAI 2008, Chicago, USA (2008)
10. Sabater, J., Sierra, C.: Reputation and social network analysis in multi-agent systems. In: Proc. of the First Int. Joint Conf. on Autonomous Agents and Multi-agent Systems, pp. 475–482. ACM Press, New York (2002)
11. Joseph, S., Sierra, C., Schorlemmer, M., Dellunde, P.: Information-Based Reputation. In: Proc. of First International Conference on Reputation: Theory and Technology (2009)
12. Sierra, C., Sabater-Mir, J.: Review on computational trust and reputation models. Artificial Intelligence Review 24(1), 33–60 (2005)
13. Neville, B., Pitt, J.: A simulation study of social agents in agent mediated e-commerce. In: Proceedings of the Seventh International Workshop on Trust in Agent Societies (2004)
14. Huynh, D., Jennings, N.R., Shadbolt, N.R.: Developing an integrated trust and reputation model for open multi-agent systems, pp. 65–74 (2004)
15. Gaur, V., Bedi, P.: Evaluating trust in agent mediated e-commerce. The International Journal of Technology, Knowledge and Society (3) 65–74
16. Huynh, T.D., Jennings, N.R., Shadbolt, N.R.: An integrated trust and reputation model for open multi-agent systems. In: Autonomous Agents and Multi-Agents systems, pp. 119–154 (2006)
17. Ramchurn, S.D., Jennings, N.R., Sierra, C.: Persuasive negotiation for autonomous agents: A rhetorical approach. In: IJCAI Workshop on Computational Models of Natural Argument, pp. 9–17. IJCAI Press (2003)
18. Esfandiari, B., Chandrasekharan, C.: On how agents make friends: Mechanisms for trust acquisition. In: Proc. of the Fourth Workshop on Deception, Fraud and Trust in Agent Societies, pp. 27–34 (2001)
19. Vasalou, A., Hopfensitz, A., Pitt, J.V.: In praise of forgiveness: Ways for repairing trust breakdowns in one-off online interactions. Int. J. Hum.-Comput. Stud. 66(6), 466–480 (2008)

20. Costantini, S., Tocchio, A.: Learning by knowledge exchange in logical agents. In: From Objects to Agents: Intelligent Systems and Pervasive Computing, Proceedings of WOA 2005 (2005) ISBN 88-371-1590-3
21. Costantini, S., Tocchio, A.: A logic programming language for multi-agent systems. In: Flesca, S., Greco, S., Leone, N., Ianni, G. (eds.) JELIA 2002. LNCS (LNAI), vol. 2424, p. 1. Springer, Heidelberg (2002)
22. Costantini, S., Tocchio, A.: The DALI logic programming agent-oriented language. In: Alferes, J.J., Leite, J. (eds.) JELIA 2004. LNCS, vol. 3229, pp. 685–688. Springer, Heidelberg (2004)
23. Sadri, F., Toni, F.: Computational logic and multiagent systems: a roadmap. Computational Logic, Special Issue on the Future Technological Roadmap of Compulog-Net (1999)
24. Beer, M., d'Inverno, M., Jennings, N., Luck, M., Preist, C.: Agents that reason and negotiate by arguing. Knowledge Engineering Review 14(3), 285–289 (1999)
25. Dung, P., Mancarella, P., Toni, F.: A dialectical procedure for sceptical, assumption-based argumentation. In: Proceedings of 1st International Conference on Computational Models of Argument (COMMA 2006). IOS Press, Amsterdam (2006)
26. Bentahar, J., Moulin, B., Meyer, J.-J.C., Chaib-draa, B.: A Modal Semantics for an Argumentation-Based Pragmatics for Agent Communication. In: Proceedings of Third International Joint Conference on Autonomous Agents and Multiagent Systems, New York, July 19-23, 2004, pp. 792–799 (2004)
27. Leue, S.: QoS Specification Based on SDL /MSC and Temporal Logic. In: van Bochmann, G., de Meer, J., Vogel, A. (eds.) Proceedings of The Montreal Workshop on Multimedia Applications and Quality of Service Verification (1994)
28. Clarke, E.M., Lerda, F.: Model Checking: Software and Beyond. Journal of Universal Computer Science 13, 639–649 (2007)
29. Milosevic, Z., Dromey, R.G.: On expressing and monitoring behaviour in contracts. In: Proceedings of The Sixth International Enterprise Distributed Object Computing Conference, pp. 3–14 (2002)
30. Bauer, A., Leucker, M., Schallhart, C.: Runtime Verification for LTL and TLTL. Technical Report TUM-I0724, Institut fur Informatik, Technische Universität München (2007)
31. Governatori, G., Milosevic, Z.: Dealing with contract violations: formalism and domain specific language. In: Proceedings of EDOC 2005, Washington, DC, USA, pp. 46–57. IEEE Computer Society Press, Los Alamitos (2005)
32. Milosevic, Z., Gibson, S., Linington, P.F., Cole, J.B., Kulkarni, S.: On design and implementation of a contract monitoring facility. In: Proceedings of The 1st IEEE Workshop on Econtracting (WEC 2004), pp. 62–70. IEEE Computer Society Press, Los Alamitos (2004)
33. Grosof, B.N., Poon, T.C.: Sweetdeal: representing agent contracts with exceptions using XML rules, ontologies, and process descriptions. In: Proceedings of The 12th International Conference on World Wide Web, pp. 340–349. ACM Press, New York (2003)
34. Dung, P.M., Thang, P.M., Toni, F.: Towards argumentation-based contract negotiation. In: Proceedings of COMMA 2008, Computational Models of Argument, vol. 172, pp. 134–146 (2008)
35. Walton, C.D.: Verifiable agent dialogues. Journal of Applied Logic 5(2), 197–213 (2007); Special Issue on Logic-Based Agent Verification
36. Wooldridge, M., Parsons, S.: Languages for Negotiation. In: Horn, W. (ed.) Proceedings of ECAI 2000, The Fourteenth European Conference on Artificial Intelligence, Berlin, Germany, pp. 393–397 (August 2000)
37. Parsons, S., Sierra, C., Jennings, N.R.: Agents that reason and negotiate by arguing. Journal of Logic and Computation 8, 261–292 (1998)

38. Dung, P.M., Thang, P.M.: Towards an Argument-based Model of Legal Doctrines in Common Law of Contracts. In: Fisher, M., Sadri, F. (eds.) Proceedings of CLIMA-IX, Int. Worksh. on Computational Logic in Multi-Agent Systems, pp. 111–126 (2008)
39. Dung, P.M., Thang, P.M., Hung, N.D.: Argumentation-based Decision Making and Negotiation in E-Business: Contracting a Land Lease for a Computer Assembly Plant. In: Fisher, M., Sadri, F. (eds.) Proceedings of CLIMA-IX, Int. Worksh. on Computational Logic in Multi-Agent Systems, pp. 91–109 (2008)
40. Brewka, G.: Dynamic argument systems: a formal model of argumentation processes based on situation calculus. Journal of logic and computation 11, 257–282 (2001)
41. Artikis, A., Sergot, M., Pitt, J.: An executable specification of an argumentation protocol. In: IJCAL 2003: Proceedings of the 9th international conference on Artificial intelligence and law, pp. 1–11. ACM, New York (2003)
42. Artikis, A., Sergot, M., Pitt, J.: Specifying norm-governed computational societies. ACM Trans. Comput. Logic 10(1), 1–42 (2009)
43. CONTRACT EU project: Contract based e-business system engineering for robust, verifiable cross-organizational business applications, 6th Framework Programme, project number FP6-034418
44. Oren, N., Panagiotidi, S., Vazquez-Salceda, J., Modgil, S., Luck, M., Miles, S.: Towards a formalisation of electronic contracting environments. In: Proc. of Coordination, Organization, Institutions and Norms in Agent Systems, the International Workshop at AAAI 2008 (2008)

Argument-Based Decision Making and Negotiation in E-Business: Contracting a Land Lease for a Computer Assembly Plant

Phan Minh Dung, Phan Minh Thang, and Nguyen Duy Hung

Department of Computer Science, Asian Institute of Technology
GPO Box 4, Klong Luang, Pathumthani 12120, Thailand
dung@cs.ait.ac.th, thangfm@ait.ac.th, nguyenduy.hung@ait.ac.th

Abstract. We describe an extensive application of argument-based decision making and negotiation to a real-world scenario in which an investor agent and an estate manager agent negotiate to lease a land for a computer assembly factory. Agents are equipped with beliefs, goals, preferences, and argument-based decision-making mechanisms taking uncertainties into account. Goals are classified as either structural or contractual. The negotiation process is divided into two phases. In the first phase, following a recently proposed framework [8] the investor agent find suitable locations based on its structural goals such as requirements about transportation; the estate manager agent determines favored tenants based on its structural goals such as requirements about resource conservation. In the second phase, we introduce a new novel argument-based negotiation protocol for agents to agree on contract to fulfill their contractual goals such as waste disposal cost.

1 Introduction

Argument-based negotiation enables agents to couple their offers with arguments, thus is believed to improve the quality of deals in such contexts as e-business, resource allocation [5]. We describe an extensive application of argument-based decision making and negotiation to a real-world scenario in which an investor agent and an estate manager agent negotiate to lease a land for a computer assembly factory.

Agents are equipped with beliefs, goals, preferences, and argument-based decision-making mechanisms taking uncertainties into account. Beliefs is structured as assumption-based argumentation framework. Goals are classified as structural if they are about static properties of purchased items or services; like a structural goal of an investor for leasing a parcel of land could be that its location is near a sea port. Goals are classified as contractual if they are about features subject to negotiation leading to the agreement of a contract; like a contractual goal for above lease is that the rental cost is lower than $.9/m^2/month$. Preferences are given by numerical rankings on goals.

The negotiation process is divided into two phases. In the first phase, following a recently proposed contract negotiation framework [8] the investor agent

M. Fisher, F. Sadri, and M. Thielscher (Eds.): CLIMA IX, LNAI 5405, pp. 154–172, 2009.

finds suitable locations based on its structural goals; the estate manager agent determines favored tenants based on its structural goals. In the second phase, agents negotiate to agree upon a contract fulfilling their contractual goals. Agents starts negotiation about a basic item or a main service. As negotiation proceeds, agents may introduce sub-items or new services to accommodate each other's needs for a better deal. For example, the estate manager offers a waste disposal service at low price to make the land lease more attractive for the investor. This kind of reward is very common in daily business. To handle this pattern of negotiation, we develop a reward-based minimal concession negotiation protocol extending the original protocol [8], which does not deal with changes of negotiated items/services during negotiation. Like its predecessor, the new protocol ensures an efficient and stable agreement.

The paper is structured as follows. Section 2.1 gives background on argument-based decision making and section 2.2 presents our new negotiation protocol. Section 3 instantiates the contract negotiation framework [8] to model the decision making of an investor (we omit the estate manager's part due to the lack of space). Section 4 is a design for implementation, and is followed by the conclusions.

2   Argument-Based Decision Making and Negotiation

2.1   Argument-Based Decision Making

An ABA framework, see [4,7,6,8,13] for details, is defined as a tuple $\langle \mathcal{L}, \mathcal{R}, \mathcal{A}, \overline{} \rangle$ where

- $(\mathcal{L}, \mathcal{R})$ is a *deductive system*, consisting of a language $\mathcal{L}$ and a set $\mathcal{R}$ of inference rules,
- $\mathcal{A} \subseteq \mathcal{L}$, referred to as the set of *assumptions*,
- $\overline{}$ is a (total) mapping from $\mathcal{A}$ into $\mathcal{L}$, where $\overline{x}$ is referred to as the *contrary* of x.

We assume that the inference rules in $\mathcal{R}$ have the syntax $l_0 \leftarrow l_1, \ldots l_n$ (for $n \geq 0$) where $l_i \in \mathcal{L}$. Assumptions in $\mathcal{A}$ do not apprear in the heads of rules in $\mathcal{R}$.

A backward deduction of a conclusion x supported by a set of premises P is a sequence of sets $S_1, \ldots, S_m$, where $S_1 = \{x\}$, $S_m = P$, and for every i, where y is the selected sentence in S_i: If y is not in P then $S_{i+1} = S_i - \{y\} \cup S$ for some inference rule of the form $y \leftarrow S \in R$. Otherwise $S_{i+1} = S_i$.

An *argument* is a (backward) deduction whose premises are all assumptions.

In order to determine whether a conclusion (set of sentences) should be drawn, a set of assumptions needs to be identified providing an "acceptable" support for the conclusion. Various notions of "acceptable" support can be formalised, using a notion of "attack" amongst sets of assumptions whereby X *attacks* Y iff for some $y \in Y$ there is an argument in favour of $\overline{y}$ supported by (a subset of) X. A set of assumptions is deemed

- *admissible*, iff it does not attack itself and it counter-attacks every set of assumptions attacking it;
- *preferred*, iff it is maximally admissible.

We will use the following terminology:

- a preferred set of assumptions is called preferred extension.
- a preferred extension of $\langle \mathcal{L}, \mathcal{R}, \mathcal{A}, \bar{\ } \rangle \cup \{a\}$, for some $a \in \mathcal{A}$, is a preferred extension of $\langle \mathcal{L}, \mathcal{R}, \mathcal{A}, \bar{\ } \rangle$ containing a.
- given a preferred extension E and some $l \in \mathcal{L}$, $E \models l$ stands for "there exists a backward deduction for l from some $E' \subseteq E$".

Agents are equipped with beliefs, goals, and preference. Following [8], an agent is defined as a tuple $< G, B, P >$, where

- $G \subseteq \mathcal{L}$ is its goal-base consisting of two disjoint subsets: $G = G^{struct} \cup G^{contr}$, where G^{struct} contains structural goals concerning the attributes of purchased items or services, for example a structural goal for leasing a parcel of land could be that its location is near a sea port; and G^{contr} contains contractual goals concerning the contractual features of purchased items or services, for example a contractual goal for above lease is that the rental cost is lower than $\$.9/m^2/month$.
- P is its preference-base mapping goals from G to the set of natural number, ranking goals according to their importance so that the higher the number assigned to a goal, the more important the goal.
- B is its belief-base represented by an ABA framework $\langle \mathcal{L}, \mathcal{R}, \mathcal{A}, \bar{\ } \rangle$, where

  - $\mathcal{R} = R_i \cup R_n \cup R_c$, where
    * R_i represents information about concrete items or services to be traded, for example the distance from a parcel of land to a sea port is 30 kms.
    * R_n consists of rules representing (defeasible) rules or norms, for example textile industries require only low skilled labour force.
    * R_c represents information related to contractual goals, for example an estate manager often offers rental discount for investors in electronics.
  - $\mathcal{A} = A_d \cup A_c \cup A_u$, where
    * A_d consists of assumptions representing items or services for transactions, for example *location*1, *location*2.
    * A_c represents control assumptions related to defeasible norms.
    * A_u contains assumptions representing the uncertainties about items or services to be traded, for example whether the labour skill available at a location is high.

A contract is viewed as a transaction between agents playing different roles, characterized by an item or service package and an assignment of values to item attributes. Formally, a contract between two agents is a tuple $< Buyer, Seller, Item, Features >$ where

- *Buyer, Seller* are different agents representing the buyer and seller in the contract
- *Item* is the item or service package to be traded in the transaction
- *Features* is an assignment of values to item/service attributes

An example contract is $< investor, estate\_manager, location2, rental = \$1.0/m^2/month >$ indicating that the estate *eatate_manager* leases *location2* to the *investor* at $\$1.0/m^2/month$.

To agree on a contract, agents engage in a two-phase negotiation process. In the first phase, the buyer agent evaluates available items or services to determine how they satisfy its needs. In the second phase, the buyer agent negotiates with the seller for items/services that have passed the first phase. Choices in the first phase are available items or services, and choices in second phase are possible deals. The value of a choice is represented by the set of goals satisfied by the choice.

Let $d \in A_d$ be a choice available to an agent $< G, B, P >$ and $g \in G^{struct}$, we says that

- g is credulously satisfied by d if there is a preferred extension E of $B \cup \{d\}$ such that $E \models g$
- g is skeptically satisfied by d if, for each preferred extension E of $B \cup \{d\}$, $E \models g$

The framework in [8] models risk-averse decision makers who consider *the value of choice d*, denoted by $Val(d)$ as the set of goals skeptically satisfied by d.

Definition 1. *Let d, d' be two choices and $s = Val(d)$, $s' = Val(d')$ be the sets of goals representing the values of d, d' respectively. Then d is preferred to d', denoted by $d \sqsupseteq d'$ iff*

- *there exists a goal g that is satisfied in s but not in s', and*
- *for each goal g', if $P(g') \geq P(g)$ and g' is satisfied in s' than g' is also satisfied in s.*

That is, choices enforcing higher-ranked goals are preferred to those enforcing lower-ranked goals.

2.2 A Reward-Based Minimal Concession Negotiation Protocol

Suppose an investor and an estate manager consider a partnership. The estate manager wants to provide not only land lease but also other estate services such as wastewater treatment. However, at the beginning the estate manager may not have full information about the investor's needs, and the investor may also not have full information about the estate services. They often start negotiation about only land lease. As negotiation proceeds, the estate manager may introduce additional services when he discovers the investor's needs. These services are called *value-added* if their values are lower than that of the main service, however they are offered at significantly lower prices than the prices that could be obtained if purchased separately from different service providers. This kind of reward is very common in business when a service provider offers an extra service at low price to increase the attractiveness of a service package. For example in our scenario the estate manager can offer waste disposal service for some kinds of industrial waste produced from the manufacturing of printed circuits when he discovers this investor's need. The reason the estate manager offers this service at low price is that he collects similar wastes from other tenants as well and then treats them in large scale. Furthermore if the investor contracts with an outside company, he has to pay extra cost for transportation.

To handle this pattern of negotiation, we extend the minimal concession protocol introduced in [8], which is itself inspired by the monotonic concession protocol in [23].

We assume that a buyer agent β needs a main service msr and a set S_β of (value-added) services. After the first phase, the buyer agent decides to start negotiation to buy msr from a seller σ and buys other services in S_β from wherever the best offers he gets. The seller σ wants to sell the main service msr, possibly packaged with other services in S_σ possibly different from S_β. A service package is defined as a set $p = \{msr\} \cup r$, where $r \subseteq S_\sigma \cap S_\beta$.

Agents negotiate to determine a concrete service package for transaction. The value of such transaction is defined by a contractual state.

Definition 2. *A contractual state is a pair $\langle p, ass \rangle$ where p is a service package and ass is an assignment of values to contractual attributes(e.g. $\{price = 10K, deliveringTime = 1week\}$). The set of all contractual states is denoted by CS while the set of all contractual states about p is denoted by CS_p.*

The preference of an agent α between contractual states can be represented as a total pre-order $\sqsupseteq_\alpha$ where $t \sqsupseteq_\alpha t'$ states that t is preferred to t' (for α). $\sqsupseteq_\alpha$ is assumed to be consistent with the partial order obtained from *Definition 1*. We assume that agent knows its preferences between contractual states. Agents are not assumed to know the preferences between contractual states of other agents except if the states have the same package. We say that: t is *strictly preferred* to t' for agent α, denoted by $t \sqsupset_\alpha t'$ if $t \sqsupseteq_\alpha t'$ and $t' \not\sqsupseteq_\alpha t$; t is *equally preferred* to t' for agent α, denoted by $t =_\alpha t'$ if $t \sqsupseteq_\alpha t'$ and $t' \sqsupseteq_\alpha t$; t *dominates* t', denoted by $t > t'$ if t is preferred to t' for both seller and buyer (i.e. $t \sqsupseteq_\beta t'$ and $t \sqsupseteq_\sigma t'$) and, for at least one of them, t is strictly preferred to t'; t is *Pareto-optimal* if it is not dominated by any other contractual state.

We also assume that each agent α possesses an evaluation function λ_α that assigns to each package p a contractual state $\lambda_\alpha(p)$ representing the reservation value of p for α. For the buyer agent β (or the seller σ, resp.), $\lambda_\beta(p)$ (or $\lambda_\sigma(p)$, resp.) is the maximal (or minimal, resp.) offer it could make (or accept, resp.). The possible deals (contracts) that agent α could accept for a package p is defined by $PD_\alpha(p) = \{t | t \in CS_p \text{ and } t \sqsupseteq_\alpha \lambda_\alpha(p)\}$. Furthermore, agents are rational in the sense that they would not accept a deal that is not Pareto-optimal. We define the *negotiation set* $NS(p)$ (about a package p) as the set of all Pareto-optional contractual states in $PD_\beta(p) \cap PD_\sigma(p)$. It is not difficult to see that for $t, t' \in NS_p$, $t' \sqsupseteq_\sigma t$ iff $t \sqsupseteq_\beta t'$, $t' \sqsupseteq_\sigma t$ iff $t \sqsupseteq_\beta t'$, and $t' =_\sigma t$ iff $t' =_\beta t$.

A package p is said to be *negotiable* if $NS(p)$ is not empty. It follows that p is negotiable iff $\lambda_\beta(p) \sqsupseteq_\sigma \lambda_\sigma(p)$ (or $\lambda_\sigma(p) \sqsupseteq_\beta \lambda_\beta(p)$).

We represent a state of a negotiation as a tuple $\langle(\sigma, v_\sigma), (\beta, v_\beta)\rangle$ where v_σ, v_β are the lastest offers of the seller agent and the buyer agent respectively. Offers are represented by contractual states. Agent starts negotiation by putting forwards its most preferred offer from the initial negotiation set $NS(\{msr\})$. That is, the seller agent offers to sell msr at $\lambda_\beta(\{msr\})$ and the buyer agent offers to buy it at $\lambda_\sigma(\{msr\})$. The negotiation state after these moves is $\langle(\sigma, \langle\{msr\}, \lambda_\beta(\{msr\})\rangle), (\beta, \langle\{msr\}, \lambda_\sigma(\{msr\})\rangle)\rangle$.

Suppose now that agents are negotiating about a package p and the current negotiation state is $\langle(\sigma, v_\sigma), (\beta, v_\beta)\rangle$. If agent $\alpha \in \{\beta, \sigma\}$ taking its turn to move next puts an offer v about the package p then v should be an element of $NS(p)$ such that $v \sqsupseteq_{\overline{\alpha}} v_\alpha$ because when an agent makes a new offer, it should be at least as preferred for its opponents (denoted by $\overline{\alpha}$) as the one it has made previously.

Instead of making new offer for package p, the agent could introduce or request a set of new services r to be included in the negotiation. To determine the asking price for the new package from the current stage of negotiation, we assume that each agent α possesses a function $f_{p,r}^\alpha : CS_p \rightarrow CS_{p \cup r}$ computing its first offer for $p \cup r$ from an offer about p, where $r \cap p = \emptyset$. It is sensible to assume that $f_{p,r}^\alpha$ satisfies following constraints

1. *Lossless.* The new offer is strictly preferred (or preferred, resp.) to its previous offer for agent α (or $\overline{\alpha}$, resp.), who introduces/requests (or who replies, resp.)
   $f_{p,r}^\alpha(v) \sqsupseteq_\alpha v$ and $f_{p,r}^\alpha(v) \sqsupseteq_{\overline{\alpha}} v$.
2. *Reward.* The seller offers r additional to p at price cheaper than the price the buyer could get r from other vendors.
   $f_{p,r}^\sigma(v) \sqsupseteq_\beta f_{p,r}(v) \sqsupseteq_\beta v$, where $f_{p,r}$ is a function returning the minimal possible cost of $p \cup r$ if the buyer purchases p at v from the seller and then purchases r from other vendors.
3. *Monotonicity.* Service inclusion retains preference order.
   If $v_2 \sqsupseteq_\alpha v_1$ then $f_{p,r}^\alpha(v_2) \sqsupseteq_\alpha f_{p,r}^\alpha(v_1)$. If $v_2 =_\alpha v_1$ then $f_{p,r}^\alpha(v_2) =_\alpha f_{p,r}^\alpha(v_1)$.
4. *Value-added.* Service inclusion expands negotiation space.
   $f_{p,r}^\alpha(v) \sqsupseteq_\alpha f_{p,r}^{\overline{\alpha}}(v)$.
5. *Flatten.* The inclusion of a set r of services can be substituted by the inclusions of individual services in r consecutively.
   $\forall asr \in r, f_{p,r}^\alpha(v) = f_{p \cup \{asr\}, r - \{asr\}}^\alpha(f_{p,\{asr\}}^\alpha(v))$; and $f_{p,\emptyset}^\alpha(v) = v$

Example 1. To motivate and explain the above constraints, let's consider a simple case where cost (say in US\$) is the only contractual attribute. A contractual state is defined by a pair $\langle p, v \rangle$ where v a natural number representing a cost of package p. So $\lambda_\alpha(p) = \langle p, \Gamma_\alpha(p)\rangle$, where $\Gamma_\alpha(p)$ is a natural number representing the reservation cost of p for α. It is reasonable to assume that $\Gamma_\beta(p \cup r) = \Gamma_\beta(p) + \Gamma_\beta(r)$ since β may have to buy each services separately from different vendors. Suppose the lowest price the seller is willing to offer r is $d(r)$, which could be considered as a fixed effective reservation price of r. Hence $\Gamma_\sigma(p \cup r) = \Gamma_\sigma(p) + d(r)$. So it is sensible for the buyer to set

$f_{p,r}^\beta(\langle p, n \rangle) = \langle p \cup r, n + d(r) \rangle$; and for the seller to set $f_{p,r}^\sigma(\langle p, n \rangle) = \langle p \cup r, n + \Gamma(r) \rangle$ where $\Gamma(r)$ is smaller than $\Gamma_0(r)$ which is the minimal possible amount the buyer has to pay for getting r on the market from other vendors (i.e. the minimal possible sum of market price of r and cost for packaging p, r together). It is sensible to expect $\Gamma_\beta(r) \geq \Gamma_0(r)$ (since β has to pay at least the minimal market price of r and packaging cost in order to obtain r from other vendors), and $\Gamma(r) > d(r)$ and $f_{p,r}(\langle p, n \rangle) = \langle p \cup r, n + \Gamma_0(r) \rangle$. It is easy to see the satisfaction of above constraints when written in simplified forms belows

- Lossless: $f_{p,r}^\beta(\langle p, n \rangle) = \langle p \cup r, n + d(r) \rangle \sqsupset_\beta \langle p, n \rangle$ and $f_{p,r}^\sigma(\langle p, n \rangle) = \langle p \cup r, n + \Gamma(r) \rangle \sqsupset_\sigma \langle p, n \rangle$.
- Reward: $f_{p,r}^\sigma(\langle p, n \rangle) = \langle p \cup r, n + \Gamma(r) \rangle \sqsupset_\beta f_{p,r}(\langle p, n \rangle) = \langle p \cup r, n + \Gamma_0(r) \rangle \sqsupset_\beta \langle p, n \rangle$
- Monotonicity: If $n_1 < n_2$ then $\langle p \cup r, n_1 + d(r) \rangle \sqsupset_\beta \langle p \cup r, n_2 + d(r) \rangle$ and $\langle p \cup r, n_2 + \Gamma(r) \rangle \sqsupset_\sigma \langle p \cup r, n_1 + \Gamma(r) \rangle$
- Value-added: $\langle p \cup r, n + d(r) \rangle \sqsupset_\beta \langle p \cup r, n + \Gamma(r) \rangle$
- Flatten: $d(r) = \sum_{asr \in r} d(\{asr\})$ and $\Gamma(r) = \sum_{asr \in r} \Gamma(\{asr\})$

After the inclusion of new services r to the current package p, the current negotiation state is changed to $\langle (\sigma, f_{p,r}^\sigma(v_\sigma)), (\beta, f_{p,r}^\beta(v_\beta)) \rangle$. From the above constraints, it follows that

- The negotiation space is changed from $\{v | v_\sigma \sqsupseteq_\sigma v \sqsupseteq_\sigma v_\beta\}$ to $\{v | f_{p,r}^\sigma(v_\sigma) \sqsupseteq_\sigma v \sqsupseteq_\sigma f_{p,r}^\beta(v_\beta)\}$, which is not empty since $f_{p,r}^\sigma(v_\sigma) \sqsupseteq_\sigma f_{p,r}^\sigma(v_\beta) \sqsupseteq_\sigma f_{p,r}^\beta(v_\beta)$.
- $\forall v \in NS(p)$, $f_{p,r}^\sigma(v) \sqsupseteq_\sigma f_{p,r}^\beta(v) \sqsupseteq_\sigma v$ and $f_{p,r}^\beta(v) \sqsupseteq_\beta f_{p,r}^\sigma(v) \sqsupseteq_\beta v$. Thus if agents could reach a deal about p then they could reach a new deal about $p \cup r$ that *dominates* the other.
- the size of $\{t | f_{p,r}^\sigma(v) \sqsupseteq_\beta t \sqsupseteq_\beta f_{p,r}(v)\}$ (or $\Gamma_0(r) - \Gamma(r)$ as in example 1) could be considered as part of a reward from the seller to the buyer.

We define reward-based monotonic concession negotiation as an interleaving sequence of concession negotiation about the package already accepted for negotiation and negotiation for service inclusion. Concession negotiation is an alternating sequence of moves between the seller agent and the buyer agent. Suppose that agents are negotiating about a current package p with the negotiation state $\langle (\sigma, v_\sigma), (\beta, v_\beta) \rangle$. A move is represented by a tuple $\langle type, \alpha, v \rangle$, where $type$ is type of the move, α is the agent making the move, and v is an element of the current negotiation space $NS(p)$. If v is strictly preferred to the agent's previous offer to its opponent ($v \sqsupseteq_{\overline{\alpha}} v_\alpha$), then $type$ is *concede*; otherwise, it is *standstill*. After a buyer's (or seller's, resp.) concession move $\langle concede, \beta, v \rangle$ (or $\langle concede, \sigma, v \rangle$, resp.) the current negotiation state is changed to $\langle (\sigma, v_\sigma), (\beta, v) \rangle$ (or $\langle (\sigma, v), (\beta, v_\beta) \rangle$, resp.). Negotiation about the inclusion of a set r of new services can be initiated with an introduction move for r of the seller agent or a request move for r of the buyer agent. An introduction move is represented by a tuple $\langle introduce, \sigma, f_{p,r}^\sigma(v_\sigma) \rangle$. A request move is represented by a tuple $\langle request, \beta, \langle p \cup r, \bot \rangle \rangle$ where $\bot$ means that the buyer is asking the seller to state its price. The buyer (or seller, resp.) will reply to the introduction (or request, resp.) move by making a reply move. If the buyer agent needs a subset $r' = r \cap S_\beta$ of introduced services, it will reply positively by making a positive reply move, represented by a tuple $\langle reply, \beta, f_{p,r'}^\beta(v_\beta) \rangle$. Similarly, if the seller agent provides a subset $r' = r \cap S_\sigma$ of requested services, it will make a positive reply move, represented by a tuple $\langle reply, \sigma, f_{p,r'}^\sigma(v_\sigma) \rangle$. A positive reply move will change the current negotiation state to $\langle (\sigma, f_{p,r'}^\sigma(v_\sigma)), (\beta, f_{p,r'}^\beta(v_\beta)) \rangle$. Agents could reply negatively by repeating its last offer to indicate that the proposal for service inclusion fails and the negotiation state remains unchanged.

Formally, a reward-based monotonic concession negotiation is a sequence $m_1, m_2, \ldots, m_n$ of alternative moves of the form $m_i =< type_i, \alpha_i, v_i >$ between a buyer agent and a seller agent where the seller agent starts the negotiation by offering to sell the main package at the buyer's reservation value, and the buyer agent replies by offering to buy it at the seller's reservation value.

Suppose now that the current negotiation state is $\langle (\sigma, v_\sigma), (\beta, v_\beta) \rangle$. Subsequent moves m_n, $n \geq 3$ could be of one of the types *introduction, request, reply, standstill, or concession,* where

- If m_n is an introduction move of the seller agent (or a request move of the buyer agent, resp.) for a set r of new services, then $m_n = \langle introduce, \sigma, f_{p,r}^\sigma(v_\sigma) \rangle$ (or $m_n = \langle request, \beta, \langle p \cup r, \bot \rangle \rangle$, resp.) where $p \cap r = \emptyset$. The current state of negotiation remains unchanged.
- If m_n is a positive reply move of the seller (or buyer, resp.) agent then the previous move is a request move of the buyer agent (or introduction move of the seller agent, resp.) for a set r of new services and $m_n = \langle reply, \sigma, f_{p,r'}^\sigma(v_\sigma) \rangle$ (or $m_n = \langle reply, \beta, f_{p,r'}^\beta(v_\beta) \rangle$, resp.) where $r' = r \cap S_\sigma$ (or $r' = r \cap S_\beta$, reps.) and $r' \neq \emptyset$. The new negotiation state is $\langle (\sigma, f_{p,r'}^\sigma(v_\sigma)), (\beta, f_{p,r'}^\beta(v_\beta)) \rangle$.
- If m_n is a negative reply move of the seller (or buyer, resp.) agent then the previous move is a request move of the buyer (or an introduction move of the seller, resp.) agent for a set r of new services and $r \cap S_\sigma = \emptyset$ (or $r \cap S_\beta = \emptyset$, resp.) and m_n, m_{n-2} coincide with exception of their types. The current negotiation state remains unchanged.
- If m_n is a standstill move then the previous move m_{n-1} is not an introduction/request move and $m_n = \langle standstill, \alpha_n, v_\alpha \rangle$.
- If m_n is a concession move then $m_n = \langle concede, \alpha_n, v_n \rangle$ and the previous move m_{n-1} is not an introduce/request move, and
  - if α_n is the seller agent then $v_n \sqsupset_\beta v_\sigma$ and the new negotiation state is $\langle (\sigma, v_n), (\beta, v_\beta) \rangle$
  - if α_n is the buyer agent then $v_n \sqsupset_\sigma v_\beta$ and the negotiation state is $\langle (\sigma, v_\sigma), (\beta, v_n) \rangle$
- A service should not be requested or introduced twice.

A seller's positive reply move or introduction move for a set r of new services where $r \cap S_\beta \neq \emptyset$ is basically an argument about a reward for the buyer agent represented in a short form. However, the move is not seen as a seller's concession since it does not suffer any loss in comparison with its previous offer.

A negotiation terminates successfully if one of the agents accepts an offer. The seller (or buyer, resp.) agent accepts an offer made in a concession move $m_n = \langle concede, \alpha_n, v_n \rangle$ by the buyer agent (or seller agent, resp.) if $v_n \sqsupseteq_\sigma v_\sigma$ (or if $v_n \sqsupseteq_\beta v_\beta$, resp.) where $\langle (\sigma, v_\sigma), (\beta, v_\beta) \rangle$ is the negotiation state after m_{n-1}.

A negotiation terminates with failure if the agents make three standstill moves consecutively. Two standstills are said to be consecutive if moves between them are only introduction, request, and negative reply moves.

Definition 3. *If a concession move leads to a successful termination, then the move is called a* finishing *move.*

Definition 4. *A contractual state t' is said to be a* minimal concession *of agent α wrt t about a package p if $t, t' \in NS(p)$ and t' is strictly preferred to t for $\overline{\alpha}$ and for each contractual state $r \in NS(p)$, if r is strictly preferred to t for $\overline{\alpha}$ then t' is preferred to r for α.*

Definition 5. *A contractual state t' is said to be a* hasty concession *of agent α wrt t about a package p if t' is a minimal concession of α wrt t about p and there exists a service $asr \notin p$ such that a minimal concession of α wrt $f^\alpha_{p,\{asr\}}(t)$ (about $p \cup \{asr\}$) is strictly preferred to $f^\alpha_{p,\{asr\}}(t')$ for α.*

So if agent α makes a minimal concession move following an introduce/request move for asr, then it will reach a state which is preferred for it to the state if it makes the introduce/request move for asr following a minimal but hasty concession move.

Definition 6. *Reward-based minimal concession negotiation is a reward-based monotonic concession negotiation where each agent only makes a minimal concession in a concession move and no agent makes a hasty concession move or a finishing move if it can make a request or introduction move. Furthermore agent standstills only if its opponent standstills in previous step.*

The following proposition shows that request and introduction moves represent a simple but effective information-seeking dialogs (for honest agents).

Proposition 1. *If both seller agent σ and buyer agent β negotiate using the reward-based minimal concession strategy, then negotiation terminates successfully by a deal containing all services in $S_\sigma \cap S_\beta$.*

Let $deal(st_\sigma, st_\beta)$ denote the deal of negotiation if σ, β use reward-based monotonic concession strategies st_σ, st_β respectively. If the negotiation terminates in failure then $deal(st_\sigma, st_\beta)$ is assigned a special value $\bot$, which is less preferred to any deal for both agents.

Proposition 2. *For any reward-based minimal concession strategy st_σ (or st_β, resp.) and any reward-based monotonic concession strategy st_β (or st_σ, resp.), there exists a reward-based minimal concession strategy st'_β (or st'_σ, resp.)such that $deal(st_\sigma, st'_\beta) \sqsupseteq_\beta deal(st_\sigma, st_\beta)$ (or $deal(st'_\sigma, st_\beta) \sqsupseteq_\sigma deal(st_\sigma, st_\beta)$,resp.).*

The following proposition shows that reward-based minimal concession strategies are equivalent.

Proposition 3. *If $st_\sigma, st_\beta, st'_\sigma, st'_\beta$ are reward-based minimal concession strategies then $deal(st_\sigma, st_\beta) = deal(st'_\sigma, st'_\beta)$.*

A strategy is said to be in *symmetric Nash equilibrium* [14] if under the assumption that if one agent uses this strategy the other agent can not do better by not using this strategy. A strategy is said to be in *symmetric subgame perfect equilibrium* [15] if for each history of negotiation h, under the assumption that one agent uses this strategy starting from h, the other agent can not do better by not using this strategy starting from h.

It is not difficult to see:

Theorem 1. *The reward-based minimal concession strategy is in symmetric Nash equilibrium and symmetric subgame perfect equilibrium.*

3   Investor's Decision Making

Foreign investors often lease serviced land plots inside industrial estate to set up factories [1,20,18]. In this session we examine how an investor in computer assembly selects a location from Vietnam industrial property market.

3.1   The Investor

Suppose an investor has analyzed computer market demand and decided to invest in assembly of low-end computers. To set up a computer assembly plant, the investor has to make decisions about technologies to be used in the plant, and location of the plant.

Goals of the investor. The investor wants to achieve

- structural goals related to technology choices, for example
  - (g_1) capacity of the plant could be easily adjusted to adapt to market demand
  - (g_2) enhancing the dynamics of assembly line (see $norm_2$ below)
- structural goals related to the location of the plant, for example
  - (g_3) qualified labour force is available at the location
  - (g_4) average wage does not exceed some threshold, e.g. $1.3/$hour$
  - (g_5) the location is near a sea port
  - (g_6) the location is eligible for sufficient government investment incentives
- contractual goals related to industrial estate services
  - (g_7) reservation price for land lease is $.9m^2/month$
  - (g_8) reservation price for waste disposal is $.3m^2/month$

The investor determines the preferences over goals by ordering them according to their importance, for example, $g_3 \sqsupseteq g_1 \sqsupseteq g_6 \sqsupseteq g_2, g_5, g_9, h_3 \sqsupseteq g_4$, and encodes the order by numerical rankings such as $P(g_1) = 5$, $P(g_2) = 3$, $P(g_3) = 6$, $P(g_4) = 1$, $P(g_5) = 3$, $P(g_6) = 4$, $P(g_7) = 3$, $P(g_8) = 3$. High ranked goals include labour and capacity adjustment. This is because computer assembly mainly concerns manual operation, so labour takes important role. Capacity needs to be adjusted according to very high expected demand variability. Lower ranked goals include wage and sea port. This is because average wage in Vietnam is very low and transportation cost is not big in comparison with the computer price.

Knowledge about technology choices. Knowledge about technologies demonstrates the technical know-how of the investor. The most important decision about technology in computer assembly concerns the structure of the assembly process. There are two kinds of assembly lines[10]. In *parallel* line, the whole assembly process is completed by a small group of workers at one workstation. In *serial* line, the assembly process is divided into sub-processes which are completed at different workstations in a specific order. The investor should know the influence of a technology choice on his goals. For example, to decide between parallel or serial lines, he should be aware of the relations between different factors:

- $norm_1$ related to g_1(capacity adjustment). If market demand changes rapidly, the investor needs to be able to *adjust production capacity* quickly. In parallel line, increasing capacity requires a duplication of *workstations*. In serial line, increasing capacity requires adding more workers to assembly line. Hence, capacity adjustment in parallel line incurs the cost of redundant workstations while in serial line it incurs the cost of modifying *working procedures*.
- $norm_2$ related to g_2(line dynamics). In serial line, workers in a workstation work under pressure from workers in other workstations of the same line. Workers at a workstation may work faster if they know that the next workstation is idle or they have just taken longer time completing the last unit, for fear that they are holding up the line. The effect of this behavior is that the line speed is maintained by workstations pushing and pulling material through the line, possibly enabling higher throughput than parallel line where no such inter-workstation pressure exists. This advantage of serial line is referred to as the *line dynamics*.
- $norm_3$ related to g_3(labour availability). Parallel line requires *higher labour force skill* than serial line. In parallel line, a worker is responsible for the assembly of the whole unit while in serial line, he is just responsible for completing tasks assigned to his workstation.
  - ($norm_{3.a}$) The investor classifies labour force skill of a location into *low* or *high*. Parallel line requires *high skill* labour force while serial line just requires *low skill* labour force[1]

---

[1] We assume a two-level classification for simplicity. The classification could be more than two.

- $(norm_{3.b})$ The investor could improve labour force skill by organizing *training programs* in electronics.

- $rule_1$(uncertainties about labour skill). If there is no information about labour skill of a location, the investor can assume that it could be *either high or low*.

The investor should know factual information about technology, for example

- $fact_1$ related to $norm_1$
  - $(fact_{1.a})$ computer assembly only requires manual tools and inexpensive general purpose workstations. The cost for factory floor for redundant workstations is not significant. Thus the cost of redundant workstations for a capacity buffer can be *ignored*.
  - $(fact_{1.b})$ changing working procedure when workers are added or removed from serial line incurs a *significant throughput loss* because the line takes sometime to stabilize.

It follows from $norm_1$ and $fact_1$ that the investor can easily adjust the capacity of the plant if he selects parallel line. However, it is costly to do so if he selects serial line.

Knowledge about locations. The investor develops a set of criteria to evaluate suitability of a location as follows

- $norm_4$ related to g_3 (labour availability). Labour availability of a location is assessed by its *population*, e.g. greater than 40K, and its labour force qualification. Labour force is qualified when its *labour force skills* meets the requirements posed by selected technology.
- $norm_5$ related to g_5 (sea port accessibility). Location should be connected to a sea port by national roads with *distance smaller than*, e.g. 35km to reduce transportation cost because some computer components need to be imported by sea.
- $norm_6$ related to g_6 (incentives). Tax reduction for at least five years is considered as an attractive incentive.

Information of land plots for lease could be as follows

- $location1$: population is 45K; distance to sea port is 30km by national road; average wage is \$1/*hour*; tax reduction in the first three years of any investment; there is an on-site electronics training center.
- $location2$: population is 46K; distance to sea port is 35km by national road; average wage is \$1.5/*hour*; tax reduction in the first eight years of any investment; the estate manager is considering building either a training center in electronics or a mansion on site; IT industry is encouraged by rental cost reduction.
- $rule_2$ (uncertainties about estate facilities). If there is no information about whether the estate manager (of $location2$) is going to develop a mansion or a training center, the investor assumes that he could develop either facility.

Decision analysis. With two candidate locations and a technology choice between serial and parallel, the investor has four options as follows: $(location1, serial)$, $(location1, parallel)$, $(location2, serial)$, $(location2, parallel)$. The satisfaction of goals wrt the above four options are summarized in Table 1. Option $(location1, parallel)$ satisfies goal g_1 because of $norm_1$ and $fact_1$; and dissatisfies g_2 because of $norm_2$. Option $(location2, parallel)$ neither satisfies nor dissatisfied goal g_3. This is because there is no information about labour skill, nor a planing to build training center or mansion of $location2$. By $rule_2$, the investor infers that estate management could build either facility. By $rule_1$, the investor can assume that the labour skill is either high or low. If the investor believes the labour skill is high, or a training center will be built, then by $norm_3$ and $norm_4$, goal g_3 will be satisfied; otherwise, g_3 will be dissatisfied. Thus, for a risk-averse agent, option $(location1, parallel)$ is the most favoured.

Table 1. Investor's goal satisfaction

Goals	Rank	Location 1		Location 2	
		Serial	Parallel	Serial	Parallel
(g_1): Capacity adjustment	5	No	Yes	No	Yes
(g_2): Line dynamics	3	Yes	No	Yes	No
(g_3): Labor	6	Yes	Yes	Yes	Undetermined
(g_4): Wage	1	Yes		No	
(g_5): Sea port	3	Yes		Yes	
(g_6): Incentives	4	No		Yes	

3.2 Formal Representation of the Investor Agent

The investor agent is represented by a triple $< G, P, B >$ where,

- goal-base $G = G_{location}^{struct} \cup G_{tech}^{struct} \cup G^{contr}$, where
  - G_{tech}^{struct} consists of structural goals related to technology choices:
    (g_1) $capacityAdjustment$; (g_2) $lineDynamics$.
  - $G_{location}^{struct}$ consists of structural goals related to location:
    (g_3) $labourAvailability$; (g_4) $wage < \$1.3/hour$; (g_5) $seaPort$
    $Accessibility$; (g_6) $incentives$.
  - G^{contr} consists of contractual goals:
    (g_7) $rental \leq \$.9/m^2/month$; (g_8) $wasteDisposal \leq \$.3/m^2/month$.
- preference-base : $P(g_1) = 5$, $P(g_2) = 3$, $P(g_3) = 6$, $P(g_4) = 1$, $P(g_5) = 3$, $P(g_6) = 4$, $P(g_7) = 3$, $P(g_8) = 3$.
- the belief-base B is an ABA framework $\langle \mathcal{L}, \mathcal{R}, \mathcal{A}, \neg \rangle$, where
  - $\mathcal{R} = R_i \cup R_n \cup R_c \cup R_f$, where
    * R_i represents information about locations

Representation of information about location 1:
$pop = 45K \leftarrow location1$; $distanceToSeaPort = 30km \leftarrow location1$;
$nationalRoad \leftarrow location1$; $wage = \$1/hour \leftarrow location1$;

$yearsOfTaxReduction(3) \leftarrow location1; trainingCenter\leftarrow location1.$
Representation of information about location 2:
$pop = 46K \leftarrow location2; distanceToSeaPort = 35km \leftarrow location2;$
$nationalRoad \leftarrow location2; wage = \$1.5/hour \leftarrow location2;$
$yearsOfTaxReduction(8) \leftarrow location2.$

* $R_n = R_n^{tech} \cup R_n^{location}$, where
  · R_n^{tech} consists of representation of norms about technologies as well as rules representing uncertainties
  Representation of $norm_1$:
  $capacityAdjustment \leftarrow parallel, asm1;$
  $difficultToAddWorkstations \leftarrow highCostForRedundant$
  $Workstations;$
  $capacityAdjustment \leftarrow serial, asm2;$
  $difficultToAddWorkers \leftarrow expensiveToChangeProcedure.$
  Representation of (the conclusion in) $norm_2$:
  $lineDynamics \leftarrow serial.$
  Representation of $norm_3$:
  $qualification \leftarrow parallel, highSkill; qualification \leftarrow serial;$
  $highSkill \leftarrow trainingCenter.$
  Representation of $rule_1$:
  $lowSkill \leftarrow notHighSkill; highSkill \leftarrow notLowSkill.$
  Representation of $rule_2$:
  $mansion \leftarrow notTrainingCenter, location2;$
  $trainingCenter \leftarrow notMansion, location2.$
  · $R_n^{location}$ consists of representation of norms about location
  Representation of $norm_4$:
  $labourAvailability \leftarrow pop > 40K, qualification.$
  Representation of $norm_5$:
  $seaPortAccessibility \leftarrow distanceToseaPort < 35km, national$
  $Road.$
  Representation of $norm_6$:
  $incentives \leftarrow yearsOfTaxReduction(X), X \geq 5.$

* R_f consists of
  Representation of $fact_1$:
  $lowCostForRedundantWorkstations \leftarrow; expensiveToChange$
  $Procedure \leftarrow.$
• $\mathcal{A} = A_d \cup A_c \cup A_u$, where

  * $A_d = A_d^{tech} \cup A_d^{location}$, where
    · $A_d^{tech} = \{serial, parallel\}$ are assumptions representing technology choices
    $\overline{serial} = parallel; \overline{parallel} = serial.$
    · $A_d^{location} = \{location1, location2\}$ are assumptions representing location choices
    $\overline{location1} = location2; \overline{location2} = location1.$

* $A_c = \{asm1, asm2\}$ are control assumptions related to norms $\overline{asm1} = difficultToAddWorkstations$; $\overline{asm2} = difficultToAdd Workers$.
* $A_u = \{notLowSkill, notHighSkill, notTrainingCenter, not Mansion\}$ are assumptions representing uncertainties about locations. $\overline{notHighSkill} = highSkill$; $\overline{notLowSkill} = lowSkill$; $\overline{notTrainingCenter} = trainingCenter$; $\overline{notMansion} = mansion$.

The investor agent's decision analysis. Table 2 shows structural goal states and their min satisfied by the composite decision. For example, g_2 is credulously satisfied by option $(location1, serial)$ -assumptions contained in the preferred extension $\{location1, serial, asm1, notLowSkill\}$. As a risk-averse decision maker, the value of an option is the min of all its goal states. So, he considers the value of option $(location2, parallel)$ is $\{g_1, g_5, g_6\}$.

Table 2. Investor's preferred extensions

Decisions	Preferred extensions		Goal states	Min
	Control asm.	Uncertainties		
(Location1,serial)	asm1	notLowSkill	g_2, g_3, g_4, g_5	g_2, g_3, g_4, g_5
(Location1,parallel)	asm1	notLowSkill	g_1, g_3, g_4, g_5	g_1, g_3, g_4, g_5
(Location2,serial)	asm1	notLowSkill,notTrainingCenter	g_2, g_3, g_5, g_6	g_2, g_3, g_5, g_6
		notLowSkill,notMansion		
		notHighSkill,notTrainingCenter		
		notHighSkill,notMansion		
(Location2,parallel)	asm1	notLowSkill,notTrainingCenter	g_1, g_3, g_5, g_6	g_1, g_5, g_6
		notLowSkill,notMansion	g_1, g_3, g_5, g_6	
		notHighSkill,notTrainingCenter	g_1, g_5, g_6	
		notHighSkill,notMansion	g_1, g_3, g_5, g_6	

4 Design for Implementation

An agent could be implemented by two separate modules. The first module is for internal decision making and the second is for bargaining. The first module is the direct translation of the agent formal representation into CaSAPI[2] and MARGO[3]. The second module is the implementation of the reward-based minimal concession strategy. A sample fragment of the seller agent's second module is as follows. We assume the agent possesses a function $f_{concede}$ to compute its next minimally conceded offer, function $noOfStandstills()$ to return the number of consecutive standstills in the negotiation, and function $notHasty(S_\sigma, p, v_\sigma, f_{concede})$ defined as $\forall asr \in S_\sigma.f^\sigma_{p,\{asr\}}(f_{concede}(v_\sigma)) = {}_\sigma f_{concede}(f^\sigma_{p,\{asr\}}(v_\sigma))$ to check if a concession is not hasty.

---

[2] www.doc.ic.ac.uk/ dg00/casapi.html
[3] http://margo.sourceforge.net/

1. The seller opens the negotiation by offering to sell the main service at the buyer's reservation value.
   $O(1, start, \beta, \{msr\}, \lambda_\sigma(s)) \leftarrow$

2. The buyer replies by offering to buy it at the seller's reservation value.
   $O(2, start, \sigma, \{msr\}, \lambda_\beta(s)) \leftarrow$

3. Suppose now that the seller has its turn at the nth move in the negotiation, S_σ contains only value-added services not introduced yet, and the negotiation after state m_{n-1} is $\langle p, (\sigma, v_\sigma), (\beta, v_\beta) \rangle$.

   (a) If the buyer standstills then the seller also standstills.
       $O(n, standstill, \sigma, p, v_\sigma) \leftarrow O(n-1, standstill, \beta, \_, \_)$

   (b) If the buyer requests a set r of services then the seller replies
       i. positively if he can provide
          $O(n, reply, \sigma, p \cup r', V) \leftarrow O(n-1, request, \beta, p \cup r, \_), r' = r \cap S_\sigma, r' \neq \{\}, V = f^\sigma_{p,r'}(v_\sigma)$
       ii. negatively otherwise.
          $O(n, reply, \sigma, p, v_\sigma) \leftarrow O(n-1, request, \beta, p \cup r, \_), r \cap S_\sigma = \{\}$

   (c) If the buyer replies or concedes then

       i. if the seller has services then
          A. he either introduces
             $O(n, introduce, \sigma, p \cup r, V) \leftarrow O(n-1, t, \beta, p, v_\beta), t \in \{reply, concede\}, r \subseteq S_\sigma, V = f^\sigma_{p,r}(v_\sigma)$
          B. or concedes provided that this is not a finishing or hasty concession move
             $O(n, concede, \sigma, p, V) \leftarrow O(n-1, t, \beta, p, v_\beta), t \in \{reply, concede\}, V = f_{concede}(v_\sigma), v_\beta \sqsupseteq_\beta V, notHasty(S_\sigma, p, v_\sigma, f_{concede})$
       ii. else, the seller concedes.
          $O(n, concede, \sigma, p, V) \leftarrow O(n-1, t, \beta, p, v_\beta), t \in \{reply, concede\}, S_\sigma = \{\}, V = f_{concede}(v_\sigma)$

   (d) The seller accepts an offer made in a concession move m_{n-1} of the buyer if $v_{n-1} \sqsupseteq_\sigma v_\sigma$
       $StopAndAccept \leftarrow O(n-1, concede, \beta, p, v_{n-1}), v_{n-1} \sqsupseteq_\sigma v_\sigma$

   (e) The negotiation terminates in failure if there are three consecutive standstills.
       $StopInFailure \leftarrow noOfStandstills() = 3$

The design of the module for bargaining of the buyer is similar.

5   Conclusion

We have extended the two-phase contract negotiation framework[8] where by in the first phase a buyer agent decides on items fulfilling its structural goals, and in the second phase it negotiates with the agent selling the item determined in the first phase to agree on a contract. The new framework improves on its predecessor by allowing agents to exchange information about each other's needs and capabilities during negotiation to change negotiated items. It also drops the assumption that the seller has no structural goals (we do not present the seller's part due to the lack of space). Our new framework, like its predecessor, allows agents to achieve Nash and subgame perfect equilibria.

The first phase is supported by a decision-making mechanism using argumentation and preferences. A number of such decision-making mechanisms exist, e.g. [11,16,12,3]. This argument-based framework can deal with decision making, uncertainties and negotiation. However, we have restricted ourself only to a simple and ideal case where we assume that the agents are honest and open to each other, and ignore the need of information-seeking in the first phase. The second phase is also supported by argumentation with only reward-based arguments. We plan to explore other types of arguments and define a communication machinery to support information-seeking in the future.

We have illustrated our approach using a scenario studied in the ARGUGRID project[4]. We believe that our approach could be fruitfully applied to scenarios where a buyer negotiates for a current item and plans possible subsequent encounters with the same seller for additional items. For example, negotiation between a car seller and buyer may cover possible after-sale services.

Several works exist on argumentation-based negotiation [17]. For example, [21] propose a protocol and a communication language for dealing with refusals in negotiation. It would be useful to see how this protocol and communication language may be used to support the two-phase negotiation framework we have defined. Also, [2] presents an abstract negotiation framework whereby agents use abstract argumentation internally and with each other. Our framework instead is tailored to the very common case in business and assumes a very concrete and structured underlying argumentation framework.

Our reward-based monotonic minimal concession strategy for fair agents is inspired by the monotonic concession protocol of [23], though it differs from it in significant ways. In our framework the agent moves alternatively where in [23] they move simultaneously. The condition for terminating the negotiation is also different. As a result, the minimal concession strategy is in symmetric subgame perfect equilibrium in our framework while the corresponding strategy in [23] is not even in symmetric Nash equilibrium. We do not use an explicit function of utilities to calculate a notion of risk to determine the player who should make the next concession as in [23] as research into practical negotiation behavior [19] shows that a strategy of making a large concession and then expects the other player to match is not a sound practical negotiation strategy as the other player often then discounts such concession as not important to the player who made it.

In this paper we have considered just two agents, not within multi-agent systems as in other existing works, e.g. [9,22] and focused instead on the full negotiation process, from the identification of issues to bargain about to the actual bargaining, thus linking argumentation-based decision making to the monotonic concession protocol.

---

[4] www.argugrid.eu

Acknowledgements

We thank the referees for constructive comments and criticisms. This work was partially funded by the Sixth Framework IST program of the European Commission under the 035200 ARGUGRID project.

References

1. Amata. Amata Vietnam (June 2008), http://www.amata.com
2. Amgoud, L., Dimopolous, Y., Moraitis, P.: A unified and general framework for argumentation-based negotiation. In: Proc. AAMAS 2007 (2007)
3. Atkinson, K., Bench-Capon, T.: Practical reasoning as presumptive argumentation using action based alternating transition systems. Artificial Intelligence 171(10-15), 855–874 (2007)
4. Bondarenko, A., Dung, P.M., Kowalski, R.A., Toni, F.: An abstract, argumentation-theoretic approach to default reasoning. Artificial Intelligence 93(1-2), 63–101 (1997)
5. Chevaleyre, Y., Dunne, P.E., Endriss, U., Lang, J., Lemaître, M., Maudet, N., Padget, J., Phelps, S., Rodríguez-aguilar, J.A., Sousa, P.: Issues in multiagent resource allocation. Informatica 30 (2006)
6. Dung, P.M.: On the acceptability of arguments and its fundamental role in non-monotonic reasoning, logic programming and n-person games. Artificial Intelligence 77, 321–357 (1995)
7. Dung, P.M., Kowalski, R.A., Toni, F.: Dialectic proof procedures for assumption-based, admissible argumentation. Artificial Intelligence 170, 114–159 (2006)
8. Dung, P.M., Thang, P.M., Toni, F.: Towards argumentation-based contract negotiation. In: COMMA 2008 (2008)
9. Endriss, U.: Monotonic concession protocols for multilateral negotiation. In: Stone, P., Weiss, G. (eds.) Proceedings of the 5th International Joint Conference on Autonomous Agents and Multiagent Systems (AAMAS 2006), pp. 392–399. ACM Press, New York (2006)
10. Furey, T.M.: Decision elements in the design of a consumer electronics assembly plant. Master's thesis, Sloan School Of Management, Massachusetts Institute of Technology (May 1999)
11. Kakas, A.C., Moraitis, P.: Argumentation based decision making for autonomous agents. In: Proc. AAMAS 2003, pp. 883–890 (2003)
12. Morge, M., Mancarella, P.: The hedgehog and the fox: An argumentation-based decision support system. In: Rahwan, I., Parsons, S., Reed, C. (eds.) Argumentation in Multi-Agent Systems. LNCS, vol. 4946, pp. 114–131. Springer, Heidelberg (2008)
13. Morge, M., Mancarella, P.: Computing assumption-based argumentation for multi-criteria decision making. Journal of Artificial Intelligence Research (January 2008)
14. Nash, J.F.: Two-person cooperative games. Econometrica 21, 128–140 (1953)
15. Osborne, M.J., Rubinstein, A.: Course in game theory. MIT Press, Cambridge (1994)
16. Rahwan, I., Amgoud, L.: An argumentation-based approach for practical reasoning. In: Proc. AAMAS 2006, pp. 347–354. ACM Press, New York (2006)
17. Rahwan, I., Ramchurn, S., Jennings, N., McBurney, P., Parsons, S., Sonenberg, L.: Argumentation-based negotiation. The Knowledge Engineering Review 18(4), 343–375 (2003)

18. Josefina Ramos, M.: Industrial estates and regional development in selected Asian countries: a review of experience. United Nations Centre for Regional Development (1991)
19. Shell, G.R.: Bargaining Negotiation Strategies for Reasonable People for Advantage. Penguin Books (1999)
20. UNIDO. Industrial Estates Principles and Practices. Technical report, United Nations Industrial Development Organization (1997)
21. van Veenen, J., Prakken, H.: A protocol for arguing about rejections in negotiation. In: Parsons, S., Maudet, N., Moraitis, P., Rahwan, I. (eds.) ArgMAS 2005. LNCS (LNAI), vol. 4049, pp. 138–153. Springer, Heidelberg (2006)
22. Zhang, D.: Reasoning about bargaining situations. In: Procs AAAI 2007, pp. 154–159 (2007)
23. Zlotkin, G., Rosenschein, J.S.: Negotiation and task sharing among autonomous agents in cooperative domains. In: Proc. IJCAI, pp. 912–917 (1989)

Author Index